THE
BIOLOGY
OF
RACE

THE
BIOLOGY
OF
RACE

JAMES C. KING

University of California Press
Berkeley   Los Angeles   London

University of California Press
Berkeley and Los Angeles, California
University of California Press, Ltd.
London, England
Copyright © 1981 by The Regents of the University of California

Library of Congress Cataloging in Publication Data
King, James C., 1904—
  The biology of race.

  Bibliography
  Includes index.
  1. Variation (Biology)   2. Race.   3. Genetics.
I. Title.
QH401.K55   1981   572   81-1345
                              AACR2

Printed in the United States of America

1 2 3 4 5 6 7 8 9

CONTENTS

PREFACE

TO THE REVISED EDITION

In the ten years since the publication of the first edition of this book many people—curious laymen as well as specialized students—have found it of value not merely in gaining insight into the problem of race but also in coming to appreciate the intimate relation between group differences and individual differences and the way in which both are related to the genetic system and the biotic program on the one hand and to cultural influences on the other. The literature on race is vast, but there is no other single work that treats the subject in the same way. I have been urged again and again to see that the book remains available for those interested.

Because developments since 1970 have made some sections of the original edition out of date, it was decided to rework the text in the light of these advances and to present a revised edition. Although numerous small changes and additions have been made throughout the work, the major alterations are in chapter 3 where recent developments in molecular biology have been incorporated and in chapter 4 where changes in general attitudes toward intelligence tests and the heritability of the I.Q. required a reworking of the discussion.

But the general focus and organization of the book are unchanged and it is hoped that this updated edition may reach an even greater number of the interested and thoughtful.

J. C. K.

PREFACE

TO THE 1971 EDITION

To a biologist the concept of race is an attempt to describe the manner in which individual variation within and between populations is related to heredity, development, and environment. It is a subject amenable to scientific investigation in which hypotheses can be formulated and tested by experiment and observation. But when applied to the human species, theories and beliefs about race are not likely to remain in the category of pure science. It is almost impossible to keep them from acquiring emotional overtones and political implications.

During the past two decades the United States has become officially committed to policies of racial integration in education and of fostering equality of civil rights and economic opportunity for all citizens. These policies constitute a clear repudiation of many practices of the past and run counter to the beliefs of large segments of the population. As a result there is much strident argument in the face of which the student and the thoughtful layman are hard pressed to find sound information on which to base opinions. Over the past twenty years the students whom I have known, both undergraduate and graduate, have had an increasing interest in the subject of race; and although the various branches of biology can provide much helpful information, nowhere has it been brought together in a concise and comprehensive account. *The Biology of Race* was written to fill this need.

Differences among human beings are in many ways similar to differences between individuals and groups within other species; so we begin with what we know of the subject in the animal world—about which it is somewhat easier to be dispassionate—and then go on to the relevance of this knowledge to the human situation. After an explanation of the concept of the biological species and its division into subspecies, there is a description of genetic units, of how they interact with each other and with the environment to produce the individual physically and to influence his behavior, and of how they are reshuffled from generation to generation; then there is a discussion of the cultural and emotional factors that have made the objective understanding of human variation difficult; and finally there is an evaluation of the unity and variety within the human species.

During its writing the manuscript was read by experts in genetics, medicine, anthropology, and educational psychology. These reviewers observed that the book would be of value to students in many different fields: biology, genetics, evolution, medicine, anthropology, sociology, education, psychology, political science, law, and philosophy. Hence an effort has been made to assume a minimum of technical knowledge on the part of the reader, and a glossary of some specialized terms has been appended for easy reference.

Many people have helped to make this book a reality. Some, like the reviewers, helped directly, and others indirectly in varying degree. To list them individually would result either in an unmanageable directory or in unjustifiable omissions. I am grateful to them all.

James C. King

New York University
School of Medicine

1

SPECIES AND SUBSPECIES

KINDS AND POPULATIONS

Ernst Mayr has described an experience he had as an ornithologist collecting in the mountains of New Guinea around 1930.

> I was all alone with a very primitive tribe of mountain Papuans, who were excellent hunters. I sent them out every morning with their guns, and for every specimen that they brought back I asked, "What do you call this one?" I recorded the scientific name in one column and the native name in another. Finally, when I had everything in the area, and when I compared the list of scientific names and the list of native names, there were 137 native names for 138 species. There were just two little greenish bush warblers for which they had only a single name (Mayr 1955, p. 5).

This story illustrates vividly an extremely important biological fact. In any locality the vast majority of living things can, like the birds of New Guinea, be classified into groups that are identifiable to anyone who takes the trouble to get acquainted with them. These groups are objective realities, recognizable by independent observers; they do not shade imperceptibly one into another. One can tell a rat from a mouse or a pigeon from a gull. Such groups are termed *species* by modern biologists, but they have been recognized by man for as long as any record has been preserved. The kinds òf beasts of the field and fowl of the air to which Adam gave names, according to the book of Genesis, were these same discrete groups.

Local hunters or primitive farmers, like the Papuan tribesmen, have never had much trouble distinguishing the local animals that had any importance for them. But as man became more sophisticated and attempted to classify animals systematically and to explain the nature and origin of the groups, he ran into some baffling problems. Some of these arose because, although the members of a species form a recognizable group, there is variation between individuals within such groups.

Hunters and farmers have always dealt with the animals in their environments in a pragmatic way. The taxonomists of the eighteenth and nineteenth

1

centuries in trying to classify animals systematically were not making on-the-spot judgments in the forest or the fence row; they worked to a great extent in leisurely deliberation on collections of preserved specimens. As a result of their educational background, most of them accepted two theoretical principles: (1) each species had been specially created at a remote time in the past and (2) every species had an ideal type of which any individual was merely an imperfect representation. The idea of special creation derived from the book of Genesis; the ideal type came from the philosophy of Plato. It may seem irrelevant what philosophical ideas were in the minds of the early taxonomists, but these two ideas—neither of which is widely held by serious scholars today—had profound and pervasive influences on animal classification; and these influences still have subtle, confusing effects on our attempts to understand animal and human variation. They have made it difficult to understand the variations that exist between individuals within a species.

Working primarily with preserved specimens, the taxonomists based their classifications on characters that did not disappear when the animals died. Size, shape, color and texture of hair, feathers or scales, size and shape of bones: in general, details of morphology were emphasized over such characters as posture, gait, feeding habits, or courtship patterns. A species was thought of as a group of animals so alike in morphology as to be representative of the same ideal type. In fact, the specimen on which the first description of a species was made was designated the "type specimen"; and every new specimen was determined by comparing it with the type. If, in the opinion of the taxonomist, the new specimen differed clearly from all known types, he established it as a new species by publishing a description of it. So the practical definition of a species came to be a group of specimens sufficiently like a given type to be classified with it.

But within every species there is individual variation. No two animals are exactly alike. The differences may be slight or striking; they may appear as a graded series or as sharply defined groups, but within a local population these individual variations will occur between parent and offspring and among members of the same litter. The swallowtail butterfly, *Papilio glaucus*, of the eastern United States is usually canary yellow with black markings. Some of the females, however, are dark brown instead of yellow. The first specimen of the species described happened to be one of the dark females. The males and the yellow females, which obviously did not match the dark brown type specimen, were described as a separate species, *Papilio turnus*. Not until it was discovered that all the sons of the dark females are yellow and that dark or light females may have dark or light daughters were both forms of the female recognized as individual variants within one species.

There is a clearwing sphinx moth, also living within the eastern United States, which in the bright sunlight resembles a bumblebee as it hovers,

2

hummingbird fashion, beside the flowers at which it feeds. These moths come out of the pupa with their wings covered with scales, but in the center of the wings the scales are sparse and weakly anchored. As soon as the moth flies, the loose scales are lost and the center of the wing becomes transparent. On the borders of the wings there are firmly anchored scales. The inner edge of this border may be straight or toothed between the veins, and three different species were described on the basis of the straightness of the inner edge of the border. Later it was shown that the contour of this border is the result of seasonal influences, probably temperature. Pupae that have lain dormant through the winter give moths with straight borders; moths emerging in late summer have toothed borders. So these different forms are merely individual seasonal variants within a single species.

Many similar cases could be cited. Sometimes males and females have little resemblance, and on occasion the two sexes have been described as separate species and treated as such until some naturalist in the field was given remedial instruction on this point by the animals themselves. The immature individuals of some species are not easily recognizable as the offspring of their parents: for example, the elver, the tiny, thin semitransparent fish that later grows into the large, thick, snakelike eel. All these examples illustrate why the notion of the ideal type made it very hard to fit the facts of individual variation into a systematic classification of animals.

But not only is there variation between individuals in a given locality; there is also variation within a species between populations in different localities. This geographical variation is not usually apparent to the local hunter, and it did not perplex the museum taxonomist so long as his specimens came from widely separated localities. But as specimens accumulated from many different collecting sites, and as more collections were made from areas between sites, it became apparent that as one moved from place to place, the sharp discontinuities between species tended to blur. Groups of specimens that had been clearly recognizable as different in the museum drawers came to be connected by intergrading forms, and often it became impossible to say, on the basis of preserved specimens, whether one was dealing with one, two, or more species or with a single geographically variable one.

In 1900 there were recognized nineteen different species of nuthatch, scattered from western Europe to eastern Siberia. But within the next few years it was discovered that between most of these there were intergrading forms, and it looked as though the attempt to classify nuthatches was impossible. In fact, about this time many taxonomists began to ask whether species existed at all.

But not all taxonomists confined their activities to museum drawers. Many were also naturalists who collected in the field and were interested in living animals. Gradually the problem shifted from that of classifying specimens to understanding the biology of the living creatures that produce the specimens. One of the most important influences in bringing about this shift was the Darwinian theory of evolution. The early classifiers had accepted the idea that species were fixed. It was their job to identify them. Darwin, in his classic *Origin of Species* (published in 1859), challenged this doctrine. He argued that species were not immutable but changed with time. He pointed out that all individual animals and plants differ from each other to a greater or lesser extent; that the potential natural increase of all creatures is vastly greater than the resources available to support them; that, as a result, natural selection operates to favor the survival of those individuals best fitted for the environment in which they find themselves; that consequently the characters of a population gradually change; and that a species may thus cease to be what it was and become something else.

The publication of the *Origin of Species* did not immediately solve the problems of the taxonomists. It set off violent arguments. Some attacked the theory as heretical and blasphemous. Others contended that it could not be reconciled with actual observation of what occurred in nature. But as time went on, more and more students of biology accepted Darwin's hypothesis. It became increasingly obvious that if Darwin was right, there should be cases where it was difficult to draw a line between species; occasional confusion could be construed as evidence of the soundness of his theory. The museum drawer concept of species gave way to a biological concept.

Mayr (1963, p. 19) has defined the biological species as "groups of actually or potentially interbreeding natural populations, which are reproductively isolated from other such groups." The emphasis is on living populations— groups of individuals who are born, grow up, reproduce, and eventually die—all of whom are part of a series of continuous cycles or generations. Males, females, young—however they may differ morphologically—are part of the same species, even though the old morphological concept sometimes stranded them in different pigeonholes.

The populations making up a species show the two types of variation already mentioned: individual and geographic. Within a local population individuals vary in sex, age, height, weight, color, and countless other characters. Some variations are discontinuous, like sex, and others are continuous, like age or height. Every population has such variation and is in this sense *polymorphic*. In some, the variety is obvious and striking—for example, the black individuals in a population of gray squirrels. In others, it is not obvious, and the general impression may be one of uniformity. Such a population is sometimes called *monomorphic*.

As one goes from a population in one locality to others nearby and to still others farther and farther away, one also finds variation between populations.

4

This is termed *geographic variation*. Nearby populations will usually show similar individual variation, both as to kinds of variants and the proportions of each. As one goes farther away, the proportions of the variants are likely to change, and one may encounter types of variants previously unobserved. At some faraway point the population may be so different from the first observed that one would scarcely classify the two as the same species if one had not seen the various intermediate forms in the areas between. Sometimes the transition is gradual and uniform over a large area; more often, there are some areas or bands where the change is more abrupt. Sometimes a whole group of characters changes together; more often, different characters replace each other in different geographical patterns, so that different local populations show different combinations. Changes with distance are likely to be gradual where population is dense and more abrupt on opposite sides of barrier regions where population is sparse or nonexistent.

Where the changes in characters are smooth and gradual from one locality to another, we say we have a *cline*, and there is no clear way of marking off one population from another. Where there are areas or bands of more abrupt change, a population centered in one area may differ rather strikingly from another in a second area. Individuals may be identified with a fair degree of certainty as coming from one or the other. Differentiated populations of this type are termed *subspecies*. A species having two or more subspecies is *polytypic*.

To achieve a satisfactory understanding of the polymorphic, polytypic biological species, one must stop trying to apply the idea of the Platonic type to living things and must grasp another concept, that of the *modal phenotype*. There is no more graphic description of the Platonic type in its most virulent form than that given by Konrad Lorenz, director for many years of the Max Planck Institute for Physiology of Behavior at Seewiesen, Germany:

> By normal we understand not the average taken from all the single cases observed, but rather the *type* constructed by evolution, which for obvious reasons is seldom to be found in a pure form; nevertheless we need this purely ideal conception of a type in order to be able to conceive the deviations from it. The zoology textbook cannot do more than describe a perfectly intact, ideal butterfly as the representative of its species, a butterfly that never exists exactly in this form because, of all the specimens found in collections, every one is in some way malformed or damaged.
>
> We are equally unable to assess the ideal construction of "normal" behavior in the Graylag Goose or in any other species, a behavior which would occur only if absolutely no interference had worked on the animal and which exists no more than does the ideal type of butterfly. People of insight *see* the ideal type of a structure or behavior, that is, they are able to separate the essentials of type from the background of little accidental imperfections (Lorenz 1966, pp. 194–95).

This mystical concept of a perfect model, never completely actualized in the crude material world, is much closer to theology than to science. It contrasts sharply with the idea of a modal phenotype envisaged by those who think in terms of the biological species. The concept of the phenotype will be discussed in greater detail in chapter 2, but here we can define it provisionally as "appearance." The modal phenotype is not a perfect abstraction like the ideal type. It is merely a recognition that in a given population the various phenotypic characters occur with definite frequencies. Any individual picked at random is very likely to possess a combination of the most frequently occurring characters. If one were to compare a population of Highland Scots with one of Ashanti, it would be obvious that for various characters such as height, weight, hair form, hair color, skin color, and the like, each population would have a mean and a distribution around the mean. For some characters, such as height, the two means would be closer together than they would be for others, such as skin color. For any number of characters selected it would be possible to say that for the Highland Scots any randomly selected individual would have some definite probability—50 percent, 60 percent, 80 percent, whatever we chose to make it—of falling within certain limits for particular characters. For the Ashanti, a corresponding statement about the same characters would, of course, have quite a different set of limits.

It is extremely difficult, laborious, and expensive to make an exhaustive metrical analysis of a large population for a large number of characters. Nevertheless, one can make approximate judgments based on whatever phenotypic data are available. This is why it is preferable to use the term "modal" rather than "average" phenotype. Many characters are difficult to measure on a precise quantitative scale, and measurements for different characters cannot be indiscriminately averaged. But if for each character a set of limits is chosen which includes the majority of the individuals, it will include the mode, the peak point of the distribution. In thinking of the modal phenotype, one must constantly picture it as an area of coincident probability and not as a sharp focal point. One single point of congruence for all characters would represent a pure type—a condition that does not exist in living things, but one in which human classifiers are constantly tempted to take refuge.

This is probably as good a place as any to point out that subspecies are in no way equivalent to breeds or strains of domestic animals. The primary difference is that breeds are not natural populations that retain the same frequencies of different phenotypes in succeeding generations when allowed to mate at random. Instead, they are artificial populations selected for characters that satisfy the whims of the breeders. These whims may be utilitarian, as for wool production in sheep; esthetic, as for vocal ability in the canary; or bizarre, as for appearance in the Chihuahua. The characters are not adaptive except as they please the breeder, and they are notoriously unstable. They will not maintain themselves from generation to generation under a system of random mating. In every generation a small number of individuals possessing the characters in approved form has to be selected from the total population to serve as the parents of the next generation. Random mating or the relaxing of selection in domestic animals almost always results in a decline in the expression of the special characters of the breed. In dogs, for example, there is a tendency toward a less distinctive, more wolflike creature. Natural subspecies, in contrast, continue from generation to generation with the same set of characters.

TWO POLYTYPIC SPECIES: JUNCOS AND MEN

A good example of a polytypic species is the North American junco. In southern New England in the late fall and winter one sees a small, slate-gray finch with a darker head and white belly and with white feathers on either side of the tail which flash like a signal when it flies. In the spring it goes north to nest in the mountains of Canada, but in the winter it comes as far south as Long Island Sound and is a common sight at feeding trays. On the coast of British Columbia, some 3,000 miles away, there is a bird of similar size, build, and habits with a black head and breast, mahogany brown back, white belly, and pinkish brown flanks. One would never think of classifying this strikingly decorated northwestern bird with the demurely colored one from New England, and they were originally described as separate species. But, in fact, they are the extreme forms of a whole complex of intergrading populations extending from Nova Scotia to southern California and from Georgia to Alaska (Miller 1941).

Within this complex, twelve different groups of populations have been named as distinguishable. The central breeding areas of these groups are outlined on the map in figure 1. A northern and mountain bird, the junco does not breed in the Mississippi Valley or on the Great Plains; but in most of the other areas between the centers of the different groups there are populations showing intermediate combinations of characters. In Pennsylvania the transition from *hyemalis* to *carolinensis* is smooth and gradual. Similarly, large transitional populations connect *oreganus*, *shufeldti* and *thurberi* along the Pacific coast. In the southwestern deserts junco populations are sparse, and the transition from *thurberi* to *dorsalis* is more abrupt, although some intermediates do occur. The transition from *dorsalis* to *caniceps* to *mearnsi* is less discontinuous. Probably the most isolated of these populations is *aikeni* in the Black Hills, which differs from all the others in having white wing bars; but there are some intergrades between *aikeni* and *mearnsi*. Over the vast expanse from New England to Alaska, there is very little obvious geographic differentiation in *hyemalis*, but in north

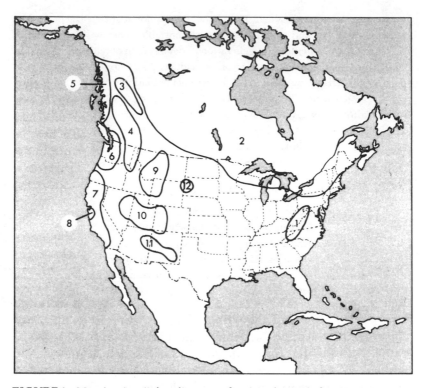

FIGURE 1. Map showing the breeding areas of twelve subspecies of the North American junco (*Junco hyemalis*), a typical polymorphic species. (1) *Junco hyemalis carolinensis*. (2) *J. hyemalis hyemalis*. (3) *J. h. cismontanus*. (4) *J. h. montanus*. (5) *J. h. oreganus*. (6) *J. h. shufeldti*. (7) *J. h. thurberi*. (8) *J. h. pinosus*. (9) *J. h. mearnsi*. (10) *J. h. caniceps*. (11) *J. h. dorsalis*. (12) *J. h. aikeni*.

central British Columbia there is a group of populations intermediate between *hyemalis* and *montanus* which has been given the separate designation *cismontanus*. These populations grade into *hyemalis* on the east and into *montanus* on the west and south. At the northern end of the Alaskan panhandle there are intermediate forms between *hyemalis* and *oreganus*.

The 1957 checklist of the American Ornithologists Union recognized four species within the complex: *aikeni*, *hyemalis*, *oreganus*, and *caniceps*. In a supplemental revision of the list in 1973 (*Auk* 90:418) *aikeni* and *oreganus* were downgraded to subspecies of *hyemalis* and *caniceps* was retained provisionally as a separate species pending further study. But since interbreeding seems to occur at all subspecific borders, these are merely transition zones and the whole complex had best be regarded as a single biological species. On the basis of this determination, the species name must be *Junco hyemalis* and for each subspecies one adds a third term: *Junco hyemalis hyemalis* for the northeastern population, *Junco hyemalis oreganus* for the north Pacific Coast groups, *Junco hyemalis dorsalis* for the birds of central Arizona, and so on for the others.

In addition to being polytypic, *Junco hyemalis* is also, of course, polymorphic. Within each local population there is individual variation, and some of these variants have characters that are more common in some other subspecies. A given specimen cannot be classified with certainty as to subspecies unless we know where it was taken. One may say of a single specimen, "It is highly probable that this is *Junco hyemalis mearnsi* from Montana," but one cannot be sure. Such an individual might come from any one of several other populations. If one has a series of half a dozen birds from one locality, the likelihood of identifying the subspecies will be much greater. The larger the sample, the greater the certainty. Populations can be identified; individuals cannot be categorized with certainty. It is also obvious, on reflection, that the number of subspecies and their boundaries are to a great extent arbitrary. *Caniceps* and *dorsalis*, for example, might be regarded as a single subspecies. Further splitting within any one subspecies might also be carried out.

In 1962 and 1963 Johnston and Selander (1964) made a careful study of several thousand specimens of house sparrows (*Passer domesticus*, also called the English sparrow) taken at twenty different points in Canada, the United States, and Mexico. These birds showed significant differences between localities in such characters as body weight, wing length, bill length and in intensity and shade of color. The sparrows show no such dramatic differences as the juncos, but this is not surprising for they have been in the United States only since 1852 when they were introduced from England and Germany, time for only 111 generations, whereas the local population of juncos have had about a thousand times as long in which to diverge since the retreat of the last continental glacier. It is interesting to know that the house sparrow has become a polytypic species in the United States in barely a century.

Another excellent example of a polytypic, polymorphic species is *Homo sapiens*, the human species. Like the juncos, human beings vary both individually and geographically. Among brothers and sisters some are tall, others short; some are slender, some stocky. Some are more heavily pigmented; some are less hairy than others. If we travel around the world, we find that the frequencies of different physical characters change with distance. If we go from Normandy to central France, the percentage of blonds decreases and the average ratio of head width to head length (the cephalic index) increases. In India, from Delhi to Madras, the proportion of individuals with heavily pigmented skin goes up.

From Omsk in Siberia to Ulan Bator in Outer Mongolia, straight, black, coarse hair becomes more common.

On the basis of their geographic variation, human populations can be grouped into subspecies just as junco populations can. Human subspecies are usually called *races*, for the subspecies concept in animal taxonomy was developed in the present century, but man has been aware of his own geographical variation for a much longer time. All peoples of whom we know anything have always recognized a distinction between themselves as a group and outsiders— Hebrews-Gentiles, Greeks-Barbarians, for example—but these distinctions emphasized social and cultural practices rather than biology. Ruth could say, "Thy people shall be my people, and thy God, my God," and be accepted. But as knowledge of the full range of human variation increased with the voyages of discovery and the later spread of Europeans over the world, attempts were made to classify the variation systematically. The grouping that is still widespread in the popular mind goes back to the work of a German physician named Blumenbach (1795) at the end of the eighteenth century. He based his classification on the study of skulls, but the five races that he recognized—Caucasian, Mongolian, Ethiopian, American, and Malay—soon became identified with the skin colors white, yellow, black, red, and brown.

But subspecies are always part of a larger interbreeding population and consequently cannot have clear and fixed boundaries. In this respect human populations are no different from those of juncos or any other animals. All the attempts during the past hundred and fifty years to bring greater refinement and precision to the delineation of human races have led only to greater and greater complication and confusion for the simple reason that the problem is insoluble. Subspecies are not discrete, isolatable units with fixed boundaries. They are merely partially differentiated pieces of one continuous unit, the species. No system of classification, no matter how clever, can give them a specificity and a separateness that they do not have.

Not only are there numerous populations that do not fit neatly into Blumenbach's fivefold system—the Polynesians, the Ainus, and the Australian Aborigines, for example—but each of the five groups can be subdivided, and subdivisions can be made to appear and disappear with complete randomness depending on what character or characters one bases them on. Furthermore, there are intergrades among the five groups: sometimes gradual clines like those between Caucasian and Mongolian extending from Europe through Siberia to Manchuria or between Caucasian and Negro from the Nile delta to the Sudan; sometimes sudden shifts as from north India into Tibet or from southern Algeria into the Niger region. Instead of accepting this situation for what it is—a complex of shifting biological equilibria—students have wasted huge amounts of energy trying to make it conform to some other pattern.

THE NATURE AND ORIGIN OF SPECIES

Although most species of animals show both individual and geographical variability, when one becomes thoroughly acquainted with living populations, one becomes aware, as did the Papuans who collected for Mayr, of a real unity within species and a discontinuity between them.

In tomato patches in the eastern United States one can find sitting on the stems of the plants and devouring the leaves large, fleshy, green caterpillars, each with a curved horn at the rear of its body. If one of these creatures is put in a jar with tomato leaves, it will go on eating and growing. When it reaches a length of about three and a half inches, it will stop eating, and within seven or eight days it will shed its caterpillar skin and become a shiny, mahogany-brown pupa—a spindle-shaped object with a projecting loop like a jug handle at one end. If the pupa is kept over winter in a cool damp place, late the following spring a large grayish hawk moth will crawl from it, leaving behind an empty shell. Its short soft wings will expand and harden and the newly emerged moth will be ready for flight.

If one observes the tomato patches carefully and collects and raises the caterpillars, one can in a year or so become completely familiar with the complicated life cycle of these insects. The moths fly at night with a swift, hovering flight like that of a hummingbird. They feed at flowers such as honeysuckle or phlox, probing for nectar with a long filamentous tongue. The jug handle on the pupa is a protective case in which the tongue develops. At dusk one can sometimes see the female moths darting from one tomato plant to another, fastening tiny, pale green eggs singly here and there to the underside of the leaves. Knowing how to recognize them, one can find the eggs by searching the underside of tomato leaves in the daytime. One becomes familiar with the tiny, newly hatched caterpillars, hardly a quarter of an inch long. One also recognizes the partly grown caterpillars, which gradually acquire the colors and the markings of the full-grown specimens.

If one makes a collection of the moths, pinning, spreading, and drying them, one comes to realize that there are two different kinds. One is grayish with a pattern of fine lines and streaks on the wings and with five orange spots along each side of the body. The other is more brownish in tone, has a pattern of mottling rather than streaking on the wings, and has one more orange spot on each side of the body: six instead of five. If one has kept careful records, one can show that the mature larvae and the pupae can also be grouped into two classes that correspond to the two kinds of moths. The caterpillars in one group are smooth and shiny and have white horizontal bands low down on the sides of their abdominal segments. These caterpillars change into pupae with large handlelike loops, extending almost two-fifths of the way along the side of the pupa. These pupae always give the grayer, five-spotted moths. Caterpillars in

11

the other group are duller, are covered with a very fine pubescence, and have no horizontal bands on their abdominal segments. Pupae from these caterpillars have shorter jug handles, not quite a third of the pupal length. These pupae always produce six-spotted moths.

One comes to realize that there are two similar but separate populations of hawk moths going through very similar life cycles. A gray, five-spotted moth always mates with another five-spotted moth and their offspring always have the larval and pupal characters of the five-spotted population. The six-spotted moths always mate with their own kind and produce six-spotted offspring. Thus the two populations, although they have very similar and parallel life cycles, are nevertheless completely separate and distinct. Each population constitutes a separate biological species, a population of interbreeding organisms going through their own life cycle over and over again.

The separation of these hawk moths into two distinct populations does not by any means remove all variation within each population. There are still many individual differences within each. Confining ourselves to the five-spotted species, we find among the adult moths differences in size, in intensity of pigmentation, in the details of the patterns on the wings. The full-grown larvae are always shiny and smooth and have the white horizontal lines on the abdominal segments, but their ground coloring may vary from light to dark blue-green through brown to black. Yet in spite of all this variation, it is clear that the five-spotted moth constitutes a unit—an interbreeding population, producing in every generation new individuals that develop in accordance with a definite pattern but within certain limits of variation and which are capable of reaching maturity and leaving offspring in spite of the vicissitudes of their environment. How are such complex developmental and ecological cycles programmed and how does a distinct cycle come into existence?

The unique specifications that make possible one of these complex programs of development, functioning, and reproduction are found coded in the DNA of the genetic material and define the genetic program of the species. The genetic program is capable of ensuring the *internal integration* of the individual; that is, it produces a structure with functional harmony between its parts and an efficient physiology capable of keeping the structure in good operating condition.

The genetic program defines the general pattern of the structure and functioning of the individuals and of the population, but it must not be envisaged as a straitjacket predetermining every detail. The program allows for individual variation within the general pattern and different programs are consistent with different degrees of variation. This is particularly true with respect to behavior. The genetic programs of the higher animals produce nervous systems capable of enormous versatility. The human brain represents the greatest achievement known in this direction.

In almost all sexually reproducing organisms, such as moths and birds and man, the specifications coded in the genetic material are provided twice. This dual or *diploid* form of the genetic material we call a *genetic complement* because in these animals a single set of specifications is not enough to direct the normal development and functioning of the individual. Nevertheless, these two sets of specifications differ from each other only in detail and constitute not two sets of different specifications but two versions of the same set. We might say that one is a paraphrase rather than a facsimile of the other. When the mature individual forms gametes (eggs or sperm), the genetic material is precisely halved so that each gamete carries one single (*haploid*) set of specifications. When two gametes combine at fertilization, the resulting *zygote* has two sets of specifications, one contributed by each gamete, and the two sets together constitute a genetic complement capable of guiding the development of the new individual and guaranteeing its internal integration according to the genetic program of the species. The genetic complement has a physical existence, the diploid set of chromosomes present in the fertilized egg and in all cells of the body that develop from it, and these two sets of chromosomes together contain two versions of the specifications of the genetic program according to which the development and the functioning of the organism proceed. Within the species the genetic program, in spite of much variation in detail, is sufficiently uniform in fundamental design for the homologous parts of the coded specifications to be interchanged as they are every time two versions, one from each parent, are brought together at the conception of a new individual.

It is this recombination of genetic material at the origin of every individual which gives unity to the species and at the same time provides for individual variation. Hence interbreeding is the critical criterion in deciding where to draw the species line. The five-spotted hawk moth constitutes a single biological species because the five-spotted hawk moths interbreed and are set off from the six-spotted moths with which they do not interbreed. The juncos of America north of Mexico constitute one biological species because, even though populations in different areas may differ strikingly in appearance, wherever these populations come into contact, interbreeding occurs and maintains the unity of the genetic program.

But a successful population has to have something more than a system providing for the development and physiological integration of its individuals. These individuals have to cope with an environment, and a given genetic system may cope efficiently or inefficiently with a given environment. An environment provides resources (food and shelter) and dangers (exposure and enemies), and the individual organisms must be programmed to take advantage of the former and avoid the latter. The five-spotted hawk moth solves the problem of nutrition by laying its eggs on tomato leaves, which are the larval food. As they sit on the underside of the leaves, the larvae are cleverly camouflaged. When fully grown, they burrow into the ground so that as pupae they lie in subterranean cells hidden from predators. The moths feed at flowers and use swift flight, camouflage, and darkness to minimize the dangers from their enemies. Juncos specialize in a diet of seeds and hence frequent semiopen country rather than deep woods or grasslands. The young are protected and cared for in a nest until they are fledged, and swift flight and camouflage are used by the adult birds against the dangers of predation.

The program that a species uses to exploit its environment may be termed its *external integration*, and the whole complex of internal and external integration together constitutes the *biotic program*. The biotic program is strongly influenced by the genetic program, but the two are not the same. Soon after its introduction into the United States in 1852, the English or house sparrow built up large populations in American cities where abundant food was available in the form of horse manure. With the disappearance of the horse from the cities, the sparrows have declined in numbers, but they have not disappeared. They are now exploiting other sources of food and have shifted their areas of greatest density of population. Although as we have seen they have developed a degree of polytypic variation over the continent, it is unlikely that there has been any profound change in the genetics of their development or physiology, but their external integration—their ecological adjustment—has changed dramatically. Their biotic program has been altered by a change in the environment. They can still exist but they are now using a different combination of their potentialities to keep going.

Sudden and major environmental changes like the disappearance of the horse from American cities in the 1920s are unusual, but they do occur. Smaller environmental shifts and fluctuations, however, are constantly occurring for all species. So the biotic program, although a real functioning system, is never rigidly fixed. Among the many sets of genetic specifications carried by all the individuals within a species, no two are exactly alike. Whenever a new individual is formed, his double set of specifications is unique. In every generation the individuals carrying certain genetic combinations leave more offspring than those carrying others. Over a period of time this produces readjustments in the genetic system. As a result, the whole biotic program is constantly undergoing change as the species and its environment interact. Over hundreds of thousands of years this kind of interaction and readjustment may result in a profound alteration of the entire biotic program. In such a case it is, of course, impossible to say whether the individuals at the end of the period would

14

or would not have been able to breed with their ancestors at the beginning of the period. After such a lapse of time, the ancestors can be known only as fossils, and the judgment as to whether the fossils represent a species distinct from the living population must be made by a paleontologist largely on the basis of morphological differences.

At some remote period in the past, two populations of the same species may have become completely separated geographically. A barrier such as a desert or an expanse of water may have developed to isolate them from each other. In such a case each population, interacting with its environment in isolation from the other, would readjust in its own way. And the environments themselves would probably undergo gradual change. As time passed, the readjustments made by the two populations would result in their becoming less and less alike. If, after a long period of isolation, such populations were again brought together by the removal of the geographic barrier, it might be that they would not interbreed but would coexist in the same area as separate populations. If this happened, we could say that the two populations were separate species, that *speciation* had occurred.

So a new species may come into being as a result of a gradual change through time of an entire species; or a part of a species isolated from the remainder may diverge sufficiently to split the former single species into two. There is some argument among the experts as to whether isolation is absolutely necessary for the splitting of one species into two, but even if on occasion the process may have occurred in some other fashion, the evidence is certainly strong that in the vast majority of cases some kind of isolation has been involved at some stage in the process. In order for two populations brought together after long isolation to live in the same area as two separate species, there are two behavioral requirements. First, there must be no interbreeding, and second, the two populations cannot follow exactly the same pattern of external integration. If two populations interbreed, the differences between them will disappear and they will become a single population. If they both try to live in the same ecological niche, eating the same food, seeking the same shelter, nesting and raising their young in the same way, the more efficient group will drive out the other, so that they will not be found in the same areas. If, however, each population has its own slightly different preferences as to food, shelter, and way of life, they are likely to coexist, each specializing in a somewhat different niche. They will then have become *sympatric species*.

The question may be raised here: What about the five- and six-spotted hawk moths? They seem to be living in the same ecological niche, and yet they exist as distinct sympatric species. Certainly, these two species do occupy very similar niches. There is evidence, however, that their biotic programs are distinctly, if subtly, different. In the first place their ranges differ. The five-spotted moth (*Manduca quinquemaculatus*) is found in Canada, the United States, Mexico, and Guatemala. The six-spotted species (*Manduca sexta*) occurs not only wherever the five-spotted hawk moth is found but also in Central America, the islands of the Caribbean, and in South America as far south as Argentina. There must be something different about the biotic program of the five-spotted species which keeps it from spreading outside North America. Where the two species occur together, they are not equally common in all localities every year. The population densities of both species fluctuate from place to place and from time to time. No one knows the exact reasons for these fluctuations. There must be differences in ecological efficiencies or one species would drive out the other. The two offer a fascinating problem in species ecology, but it would be a complex one to solve. One ecological fact that may help the two species to coexist is their differential susceptibility to parasites. The larvae of hawk moths are attacked by parasitic wasps and flies that lay eggs in or on the caterpillars. Parasitic grubs hatch from these eggs and feed on the tissues of the caterpillars, ultimately killing them. In the wild state many caterpillars of both species fall victim to these parasites. Some species of parasites attack both species of hawk moths, but most species of parasites have a distinct preference for either the five-spotted or the six-spotted moth. If the parasite population attacking one species of hawk moth builds to a high density in a given locality, it may very greatly reduce the population of that hawk moth and have little effect on the population of the other. This may very well contribute to the complementary fluctuation often observed. The difference in susceptibility to parasites is certainly not the sole reason why these two species of hawk moths can coexist, but it is probably contributory, and it illustrates how complex the biotic program can be, influenced by the preferences of bizarre predators.

Failure of two populations to interbreed is commonly termed *reproductive isolation*. This does not necessarily mean that matings between the two groups are sterile. Interbreeding may be prevented by other factors. The two populations may frequent different habitats and hence rarely meet. They may have different courtship patterns, so that signals of sexual advance and response are not mutually understood and do not lead to a sequence resulting in copulation. There are many cases where fertile matings between species can be produced under artificial conditions, but where the two exist side by side in nature without the production of hybrid offspring. Of course, sterility, where it exists, is the ultimate mechanism of isolation. But whether it exists or not, behavioral factors in the broadest sense form the most usual basis for the prevention not only of hybridization but of interspecific dalliance.

One can easily see that subspecies are potential new species. A local population that has arrived at its own range and proportions of individual variation which set it off from other populations may, if isolated, very well continue to become more different. If the isolation continues long enough, the isolated subspecies may diverge so strongly that it may become reproductively isolated from the remainder of the species. Members of the two populations may then meet and live in the same area without interbreeding. But all subspecies are not destined to become separate species. Some go on being subspecies; some merge with neighboring populations, merely shifting the subspecies complex; others become extinct.

Since subspecies interbreed where they come together, it is impossible by definition to have two subspecies of the same species in the same locality—to have *sympatric subspecies*. It has been argued by some taxonomists that in addition to geographical subspecies there are also ecological ones that are sometimes sympatric. It is certainly true that different subspecies are often ecologically distinct. In one part of its range an animal may live in marshes; in another part it may be found in well-drained woods. Many species of birds and animals live both in lowlands and at high elevations. But the preponderance of evidence is that where ecologically different populations of the same species actually come together, they interbreed; if they remain distinct, they do so because in fact they are spatially separated. The individuals do not meet each other in the ordinary course of their activities. Truly sympatric ecological species do not seem to exist.

The one animal that comes nearer than any other to having sympatric subspecies is man. Social classes are populations that are to a degree ecologically distinct, and it is generally true that human mating is more likely to occur within class lines than across them. This has the effect of decreasing the randomness of human mating within a locality and of preserving any genetic differences there may be between classes.

Social classes are never formed by dividing a population in such a way as to give genetic identity to the two groups. When an upper class is established by conquest, the likelihood of its being genetically distinct is easy to conceive. When a working class is built up during a period of industrial development, its recruits are generally attracted from outside the community and often represent people of somewhat different origin from the managerial and administrative class. In northern American cities in the period up to World War I the working classes were recruited largely from European immigrants. These immigrants were not in most cases genetically remote from the older American population, but the new and the old groups did not by the fact of their presence in the same community achieve genetic equilibrium. Where classes in the same area have been recruited from genetically dissimilar elements, class structure has the effect of slowing down the natural tendency of the whole population to come to equilibrium. But there seems to be no historical evidence that class barriers have ever been rigid enough to cause or permit an increase in genetic divergence between classes. Classes are no more likely than geographical races to lead to speciation in man, but, as we shall see later, confusion concerning the relation of race and class and man's emotional reactions to both vastly complicate the problem of trying to view human variation with objectivity.

Since the various populations of all species are constantly undergoing complex changes, it seems reasonable to predict that now and then one should encounter the untidy situation of not being able to decide whether two populations are one species or two. *Allopatric species* are one example. Two groups of populations that are morphologically and ecologically similar live in two widely separated areas. They do not interbreed because they never meet. If they came together, would they amalgamate or would they remain specifically distinct? In such cases the taxonomist can do little beyond making a best judgment based on the degree of morphological, behavioral, and ecological differences. Even an experimental mating between captured specimens would not answer the question definitively.

An even more interesting and anomalous situation is what is known as *circular overlap*. Extending from Great Britain to Scandinavia and along the Arctic coasts of European and Asiatic Russia are various populations of herring gull of the genus *Larus*. These populations interbreed where they come together. The populations of eastern Siberia interbreed with others in Alaska and these in turn interbreed with still others along the Arctic coasts of North America. The population of eastern North America, *Larus argentatus*, has crossed the Atlantic to Great Britain and Scandinavia, where it coexists without interbreeding with the European population *Larus fuscus*, even though there is evidence of continuity of interbreeding between the contiguous intervening populations. Subspecies at the extreme ends of a very wide range have diverged sufficiently to behave like good species where they come together. Some two dozen similar cases of circular overlap are known, including not only birds but butterflies, rodents, bees, and salamanders.

The problem of evaluating the specific status of allopatric populations and the impossibility of neatly classifying the number of species involved in a case of circular overlap are embarrassing to one who demands a world capable of tidy, static classification; but to one who has an understanding of the complex dynamic process of population change and species formation, they are exciting examples of situations to be expected. And it still remains true that in any given locality species are distinct and do not melt into each other.

2

GENETIC INFORMATION AND THE BIOTIC PROGRAM

We have seen that a species has a common recipe, or set of specifications, which defines its biotic program, guides its development, physiology, behavior, and reproduction, and causes these to fit harmoniously into the environment. That every individual gets his own set of specifications in two versions—one from each parent—knits the entire species into a single fabric and at the same time allows for individual and geographic variation. This unity and continuity of the living species has been recognized by biologists and naturalists for many decades, but the details of just how the complex process could be explained were simply not available. Today we are beginning to achieve some understanding of this unity and continuity of species in terms of molecular biology. Our insight is being enhanced by recent developments not only in genetics and biochemistry but also in two specialized fields of engineering: information theory and cybernetics.

The fundamental unit on which the concept of the species is built is, of course, the individual. Species are groups of populations, but the populations are made up of individuals, which function as units. Any inferences we make about populations or species must be based on observations on individuals. The problem is, what causes the individuals to function and behave as they do, and what is the nexus between one generation and its successor that keeps the new individuals's development and behavior consistent with the continuity of the species? In seeking an answer let us examine first the relation between the individual and his set of genetic specifications, between his phenotype and his genotype; second, the way in which cybernetics and information theory can help us understand biological processes; third, the way in which genetic specifications can produce continuous variation; and finally the model of polygenic inheritance.

PHENOTYPE AND GENOTYPE

We identify the people we know by their individual characteristics. Some of these, such as height, coloring, body build, are physical; others, such as posture, voice, and facial expression, are behavioral. Still others, of which we are normally not aware, such as blood group or reaction to a certain drug, are chemical. The term *phenotype* is used to refer to the combination of observable traits possessed by an individual. In addition to his phenotype, an individual also has a *genotype*, composed of the genetic specifications inherited from his parents. Sometimes we can make valid inferences about the genotype directly from the phenotype. If a human individual is phenotypically male, we can conclude that he inherited a Y chromosome from his father. If he has blue eyes, we can be reasonably sure that he has inherited one allele for blue eyes from each parent and that consequently he carries no allele for brown eyes. But we cannot observe the genotype directly; we must infer it from the results it has produced during the development of the individual. At any given point in his life the individual's phenotype is the result of the interaction between his genotype and the environmental influences to which he has been up to that time subjected.

In *Drosophila melanogaster*, the vinegar fly, there is a recessive allele that, when homozygous—inherited from both parents—produces flies with yellow rather than brown bodies. When these yellow flies mate they produce yellow offspring. Such flies are phenotypically and genotypically yellow. Ordinary wild-type flies normally produce offspring with brown bodies, but if we take their eggs and raise the larvae on food containing silver salts, they develop into flies with yellow bodies like the homozygous recessive yellows. If we take eggs from these flies and raise the larvae on food without silver salts, they will develop into wild-type flies with brown bodies. The silver in the food interacts with the wild genotype to produce a phenotype like that of the homozygous recessive mutant, but it does not change the genotype, which is passed on in unaltered form to the offspring. The flies made yellow by the silver salts in their food are said to be yellow *phenocopies*: environmental influences have given them a phenotype not characteristic of their genotype under other environmental conditions. Phenocopies involving many kinds of characters are widely known among all sorts of organisms. They exist among human beings. Very dramatic illustrations are cataracts and deafness produced by *rubella* virus acting on a fetus in the early months of gestation and the suppression of the long bones of the arms and the legs of the embryo by the sedative Thalidomide taken by the mother during the sixth or seventh week of pregnancy. These deformities resemble phenotypes produced by rare genetic defects.

The pigmentation of the human skin is a character that shows quite clearly the complexity of the relation between genotype and phenotype. Among populations of European origin there is very considerable individual variation in the color of the skin. Some people are pale, others swarthy; and there are all sorts of gradations in between. There appears to be a genetic component in skin pigmentation among Europeans. Color is often similar between parent and offspring; and offspring of dissimilar parents are often intermediate. But the presence of a continuous series of intergrades strongly suggests that skin color is not controlled by a simple set of genetic factors. There is a recessive allele that, when homozygous, produces an albino—a person with no pigment whatever. This allele has been found in many pedigrees and behaves as a simple Mendelian recessive. But it is very rare and its frequency has nothing to do with the range of pigmentation found in the general population.

But superimposed on whatever genetic influence there may be on skin color is a strong environmental effect. The skin of most individuals darkens considerably as a result of continuous exposure to sunlight. Such exposure can transform a pale individual into a swarthy one. Among Europeans, therefore, phenotypic skin color is not a reliable index of genotype.

In populations outside Europe, skin color is, on the whole, somewhat darker, ranging from pigmentations very near the European phenotypes in North Africa, western Asia, and northern China to the extremely heavy pigmentation found in the populations of central Africa. Among the light individuals, wherever found, exposure to sunlight will have the effect of increasing pigmentation; but this effect practically disappears among the most heavily pigmented. Whenever matings occur between persons of different levels of pigmentation, their offspring tend to be intermediate; and when these intermediates interbreed, their progeny tend to vary between the extremes of the original parents.

CYBERNETICS AND INFORMATION THEORY

A living organism is a self-developing, self-reproducing cybernetic device. A *cybernetic device* is one that follows a set of operations through a cycle by means of a system of feedbacks effected by the transmission of information from component to component within itself. One simple example is an automatic heating system composed of a heater and a thermostat. When the temperature is below a certain level, the thermostat closes a switch that activates the heater. The heater runs until the temperature rises to a given point, the thermostat then opens the switch, and the heater stops running. Cybernetic devices of enormously greater complexity are now familiar to everyone, in their achievements if not in their functioning: computers able to solve complex equations, automatic industrial plants for refining petroleum or making chocolate cake, and craft able to maneuver and navigate in space and send back information on schedule or at command. But all cybernetic devices operate by means of the communication of information—signals or messages—from one component of the device to

ages of varying size known as *chromosomes*. Since every nucleotide pair constitutes two bits of information, *E. coli* has 7.6×10^6 bits of information in its genetic complement. You and I each have 1.42×10^{10}.

In an organism messages are generally molecules synthesized within it or brought in from the environment. The channels are mechanical transport, diffusion, and molecular attraction. Redundancy is high, clearly shown by the resilience of individuals and populations. It exists in many forms: diploidy in many plants and animals, for example, the presence of the genetic specifications in two parallel versions, or in the maintenance of enzyme levels high enough so that a normal allele can do the work alone when paired with a recessive mutant. In many plants and animals inbreeding seems to cut down the efficiency of metabolic functioning, suggesting that outbred individuals have a higher level of redundancy in their genetic information (King 1961).

Just as a set of engineering or architectural specifications is necessary to direct the workmen and the automatic machines that produce an automobile, a refrigerator, or an automatic pilot, a set of genetic specifications is necessary to direct the biotic program of every organism, to determine the limits of its structure and behavior at every point in its life cycle. Blueprints or taped instructions do not by themselves build houses or construct appliances. They have to exist in an environment where they stimulate the interaction of administrators, workmen, machines, and materials. Similarly, genetic information coded in DNA will not produce an organism unless it is furnished with the proper environment.

Directions for the production of this proper environment are coded in the information in the genetic complement and, during the life cycle, the female produces an egg that constitutes the necessary environment for the start of a new individual. Within this specified environment she leaves one half of a genetic complement. The male gamete—the sperm—brings in the other half, and development of the new individual then starts. The critical information found in the cytoplasm of the egg is, in general, species specific. In some frogs and salamanders it is possible to remove the genetic material from one egg and substitute the genetic material from another. This procedure carried out within the same species may initiate normal development, but when a genetic complement of one species is transplanted into cytoplasm of another, development usually fails. In *Drosophila melanogaster* there is a recessive allele called deep orange—because it gives that color to the eyes—which causes the homozygous females to produce eggs with a deficiency in the cytoplasm that will not allow the development of the homozygous deep orange genotype (Counce 1956). Such females mated to deep orange males are sterile. Homozygous deep orange flies can be obtained only from *heterozygous* females in which there is one wild-type and one deep orange allele at the deep orange locus. The one wild-type allele produces cytoplasm that will allow the homozygous deep orange genotype to develop. In this case the same genotype is lethal or viable depending on the cytoplasm in which it comes to lie.

Not only is a genetic complement dependent on information in the cytoplasm of the egg for its start in guiding the biotic program through a new cycle, but as development proceeds, the organism becomes dependent on the external environment for further information. Substances are taken in to provide energy for the metabolic processes. Energy itself is a form of information; in a completely random universe there would be no free energy. Other raw materials are taken in for the synthesis of the substances and structures of the growing organism. These also are information provided by the environment. The information in the genetic complement controls the methods of finding these substances, getting them into the organism, and directing what to do with them once they are in.

If the substances that the genetic program expects from the environment are not there, the phenotype will be altered, resulting in death in cases of extreme deprivation, in stunted growth when nutritional deficiency is less drastic, or in deformities when there is a deficiency of a specific substance required by the program—such as vitamin D in the case of rickets. An unusual substance or a usual substance in unusually high concentration may modify the phenotype. An example of the first is the development of yellow bodies by *Drosophila* raised on food containing silver. An example of the second is the flecking of the enamel of the teeth of persons who have grown up where the natural drinking water had an abnormally high concentration of fluoride. If the environment furnishes substances incompatible with the genetic program—which give false signals to the metabolic components—these act as poisons injuring or killing the organism.

And the total environment must lie within certain energy levels. If the temperature is too high or too low, the biotic program will come to a halt. So a genetic complement is a body of information that in a favorable environment, will take free energy from the surroundings and organize it according to a continuing cyclical pattern, the biotic program.

The information in the genetic complements of the members of a population is used over and over, generation after generation, and is sometimes referred to as the *gene pool*. A new individual gets his genetic complement by having half a complement drawn at random from the full complement of each of his two parents. The genetic information in the individual is his genotype. The information in the gene pool of the whole population determines the genetic program, or system, of the population. In spite of individual differences in detail, all individuals in the population have the same genetic system. In fact, all the individuals in all the populations constituting a species have the same genetic system. Analogous parts of the genetic information can be interchanged throughout the whole group of populations. This is what makes them a species.

A crude but perhaps helpful comparison is one between a genetic complement and a movie film. A film with a sound track composed of silver grains designed to produce sound by activating a photosensitive component will give both picture and sound when run in a projector with the appropriate components. But the same film run in a projector designed for a sound track containing magnetic particles rather than silver grains would give no sound reproduction at all. We might say that there are two species of sound reproduction. Within a biological species the genetic material is all coded in a manner so similar basically that whenever a full complement is put in the proper setting, the probability is high that the result will be the development of an individual of the modal phenotype with effective morphology and efficient physiology (internal integration) and competence in dealing with the environment (external integration)—a successful unit in the biotic program.

THE GENETICS OF CONTINUOUS VARIATION

A series of experiments carried out some years ago by C. H. Waddington (1953) throws much light on the nature of the kind of genetic system that produces a complex array of phenotypes like those of human skin color. The experiments, however, were not done on humans but on the old standby for genetic investigation, *Drosophila melanogaster*. The wing of these flies has a short crossvein connecting the third and fourth longitudinal veins. In natural populations one finds, once in a great while, a fly in which this crossvein is interrupted or even lacking completely. Waddington discovered that if pupae were subjected to a slight rise in temperature at a given period in their development, some of the emerging flies would have reduced or obliterated crossveins. So he selected flies in which heat treatment had obliterated the crossvein and subjected their offspring to heat treatment in the pupal stage. The result was an increase in the proportion of flies with the crossveinless phenotype. This process of heat treatment and selection was continued and after seventeen generations about 90 percent of the heat-treated pupae gave crossveinless flies. In every generation some pupae, sibs of those given treatment, had been set aside and given no heat treatment. All the flies from these untreated pupae had complete crossveins until generation 14, when a few crossveinless flies appeared from the untreated pupae. These crossveinless flies were interbred and there were some crossveinless flies among their offspring. These were used to found lines that were given no further heat treatment. Selection was carried on for several more generations, and lines were established which gave from 90 to 100 percent crossveinless flies with no heat stimulus at all.

With *Drosophila melanogaster*, by using chromosomes carrying mutant markers it is possible to analyze the genetic system of a population and find out where in the chromosome set the information associated with a given phenotype is located. When the crossveinless line was analyzed, it turned out that every one of the chromosomes contained elements that contributed to the likelihood of the appearance of the crossveinless phenotype. The more chromosomes from the crossveinless line a fly carried, the more likely it was to be crossveinless and the more likely was the expression of the phenotype to be extreme—with the crossvein obliterated rather than merely interrupted.

A fly failing to develop a crossvein because it was overheated as a pupa and a man getting a tan from lying on a beach in the sun may not seem to have much similarity. But there is probably a fundamental parallel between the genetic and the developmental processes involved. The loss of a crossvein as a result of heat treatment is a response to an environmental stimulus. Selection of individuals who have responded in this way is selection of those genotypes with greater likelihood of response. Interbreeding such genotypes produces some offspring still more likely to respond; and gradually selection builds up genotypes so capable of response that they respond spontaneously without the environmental stimulus. Further selection of the spontaneously responsive genotypes can finally establish the response as normal for the population, making it part of the modal phenotype. This process must not be confused with the inheritance of acquired characters. It was not the undeveloped crossvein in the fly which made its offspring likely to develop the same way. The lack of the crossvein made it possible to identify the fly as having a genotype that would be likely to produce responsive offspring. Without selection, heat treatment for many generations would not be at all likely to increase the proportion of responsive individuals.

In the case of human pigmentation it is quite likely that increased ability to tan—to respond to environmental stimulation—is correlated to some extent with a tendency to produce pigment spontaneously. Even though this correlation is not high, it may very well be that selection for efficient tanners, or against inefficient ones, increases the likelihood of spontaneous pigment production and, if carried on indefinitely, may produce extreme pigmentation in the absence of environmental stimulus. The difficulty of being sure whether this is the explanation for racial differences in pigmentation is that we do not know precisely what the adaptive significance of pigmentation is. We shall return to this point when we consider problems of human adaptation.

There are other illustrations of a connection between ability to respond to the environment and the development of a spontaneous response which make it reasonable to believe that the mechanism is a common one. The ostrich is a heavy bird, and in its relaxed moments it relieves the strain on its leg muscles by resting its body on the ground. The underside of the body touches the ground at two points, one in front, the other behind. At these points hard calluses develop. These calluses constitute a response to environmental stimulus. But before the ostrich hatches, the spots where these calluses will form already show a thickening of skin in the embryo. Selection for the ability to form a callus has resulted in a genotype so capable of callus formation that it appears to act impatiently in anticipation of the coming stimulus. A very similar state of affairs occurs on the sole of the human foot. The thickening of the skin that develops in the growing child as a result of pressure of the foot against the ground actually begins before birth.

THE POLYGENIC MODEL

Mendelian inheritance is usually described as the process by which one of two alternate alleles of a gene is inherited from each parent so that the offspring has one of the following combinations: two dominant alleles, two recessive alleles, or one of each. If both parents are heterozygous, that is, have one of each allele, the three genotypes will occur among the offspring with the following probabilities: one homozygous dominant, two heterozygotes, and one homozygous recessive. Phenotypically the two heterozygotes will be indistinguishable from the homozygous dominant. This results in the famous 3:1 ratio. In many human families blue eyes behave according to this model, as a character recessive to dominant brown. In many pedigrees it is possible to deduce from their ancestry and their progeny the genotypes of many individuals, those with blue eyes being homozygous recessive and the brown-eyed ones either homozygous dominant or heterozygous depending on the presence or absence of blue-eyed ancestors or descendants. In many kinds of animals cases are known of this type of inheritance—a given phenotype characteristic of the homozygous recessive and another phenotype expressed in the homozygous dominant and the heterozygote. Single comb recessive to rose comb in chickens, black coat recessive to wild agouti coat in mice, scarlet eyes recessive to wild-type brick red in *Drosophila*: these characters and many similar ones were discovered by the early Mendelians and used in their demonstrations and analyses of the heredity of unit characters. In fact, there was a strong tendency to exaggerate the importance of such contrasting characters and to think of the individual as little more than a large list of unit characters pasted together. That this erroneous conception has not been permanently laid to rest is shown by the following sentence from a discussion of genetic defects by Walter Sullivan in the *New York Times* of June 13, 1970: "Each individual carries two genes, or bits of genetic instruction, for every human characteristic: one from each parent."

But there is much more to heredity and development than a mere aggregation of unit characters. This was realized by geneticists from the beginning, even though some of them were sometimes carried away by the mystique of genetic ratios. Most characters that account for individual differences in animals and in humans do not exist in two contrasting states. Instead, they vary continuously from very little to very much, and the offspring of a given mating do not fall into clearly demarcated groups with fixed ratios between them. This is true of such characters as height, weight, body proportions, and, as we have already pointed out, pigmentation.

The heredity of attributes that vary continuously is controlled by *polygenic systems*. If instead of thinking of one pair of alleles, we think of several pairs, each existing in a plus or minus form, we can construct a model of how such a system may work. Suppose we think of stature as controlled by four independently segregating loci at each of which there are plus or minus alleles, and suppose that every plus allele adds one increment of stature equal to that added by every other plus allele. Under such a system (see table 1) an individual at one extreme with the minimum height will have eight minus alleles—two at each locus, four inherited from his mother and four from his father. At the other extreme there will be individuals with eight plus alleles, taller by eight of the equal increments. Between these extremes will be the genotypes having all the intermediate combinations from only one plus allele to seven plus alleles. There will be nine combinations in all.

Among a large number of offspring of parents that are both heterozygous at all four loci, the nine possible genotypes will occur in the proportions shown in table 1. The extreme genotypes, zero plus and eight plus, will each occur only once in 256. These proportions derive from the coefficients of the nine terms in the expansion of the binomial $(a + b)^8$. The numbers in the numerators (the coefficients) approximate a normal distribution with the mean, mode, and median at the four plus, four minus genotype.

In table 1 the genotypes are classified according to the number of plus

TABLE 1. Model of a Polygenic System for Stature
There are four independent
loci, at each of which a plus
or a minus allele is possible.

Number of plus alleles[1]	0	1	2	3	4	5	6	7	8
Probability of each genotype[2]	$\frac{1}{256}$	$\frac{8}{256}$	$\frac{28}{256}$	$\frac{56}{256}$	$\frac{70}{256}$	$\frac{56}{256}$	$\frac{28}{256}$	$\frac{8}{256}$	$\frac{1}{256}$

[1] Each plus allele contributes one unit of stature above a minimum.
[2] The mating is between two quadruple heterozygotes.

alleles regardless of which loci they occupy. But individuals with the same number of plus alleles, although they will be phenotypically alike, will give different types and proportions of offspring depending on which loci carry the plus alleles. Suppose we mate two individuals, each with two plus alleles at locus 4 (see table 2). All offspring from such a mating will be phenotypically and genotypically like the parents. All will have two plus alleles, both at locus 4, one received from the mother and one from the father. Neither the mother nor the father has any other plus allele to pass on. But another mating between two individuals each with two plus alleles at different genetic loci, and hence phenotypically like the parents in the first, will produce offspring varying from zero to four plus alleles, spanning one-half the distribution both phenotypically and genotypically. In this second mating one parent is heterozygous at loci 1 and 2, the other at loci 3 and 4. One offspring may inherit no plus allele from either parent; another may inherit two from each. Still others may inherit one from one parent and none from the other (total 1), one from each (total 2), or two from one and one from the other (total 3).

From such a model it is easy to see how, even though a character is under genetic control, it may vary continuously, be normally distributed, and segregate differently among the offspring of different matings between phenotypically similar parents. But, in fact, this model is vastly oversimplified in at least three ways. In the first place there must be extremely few, if any, quantitative characters controlled by as few as four genetic loci. Increasing the number of loci greatly increases the number of genotypes and smooths out the distribution curve. Second, although an individual may have only two different alleles at a single locus, in the whole population the alleles for a given locus almost certainly represent a graded series of many different alleles with varying functional efficiencies. In any one individual any two of these may be paired. Finally, every genotype is susceptible to environmental modification so that the phenotypes spill over into adjacent genotypic boxes, and the degree of spilling is not a fixed value predictable on a formula for genotype-environment interaction, but a set of varying probabilities. In table 1, for example, the phenotype of the individual with three plus alleles will be most likely to show three increments of increased height over the shortest phenotype. But some three plus genotypes will show less and some more than this most likely figure. As one goes farther away from this most likely phenotype, one will find fewer, but some, three plus genotypes that through accidents of internal malfunctioning or unusual environmental pressures have undergone a development that has left them with an improbable phenotype.

This model of a polygenic system is, of course, only a hypothesis. We have no ironclad proof of a large number of genes controlling human stature or skin

TABLE 2. Offspring Differences from Matings of Identical Phenotypes

MATING 1: BOTH PARENTS HOMOZYGOUS AT THE SAME LOCI

Loci	1 2 3 4		1 2 3 4
Parents	− − − ‡	×	− − − ‡
	− − − +		− − − ‡

Offspring	− − − +
genotype	− − − +

MATING 2: BOTH PARENTS HETEROZYGOUS AT TWO LOCI, BUT NOT THE SAME TWO

Loci	1 2 3 4		1 2 3 4
Parents	− − + +	×	+ + − −
	− − + −		+ + − −

Possible offspring genotypes

```
− − − −    − − − −    − − − −    − − − −    − − − −    − − − −
− − − −      −        − − + −    − − + +    − + + +      + +
          + +
                    − − − −    − − − −    − − − −
                    − − + −    − + + −    + + + −

                    − − − −    − − − −    − − − −
                    − + − −    + + − −    + + − +

                    − − − −    − − − −    − − − −
                    + − − −    + − + −    + − + +

                               − − − −
                               + − − +

                               − − − −
                               − + − +
```

pigmentation. We do not know just how many there are, nor can we say where they are located among the 23 pairs of human chromosomes. But the evidence for the polygenic model comes from so many sources and has made possible the accurate prediction of the results of so many genetic experiments and practical plant and animal breeding programs that in its broad outlines it is accepted by practically all geneticists.

A very convincing demonstration of the control of a continuously varying trait by numerous genes of additive effect scattered throughout the chromosome set has been made for resistance to DDT in *Drosophila melanogaster* (King and Sømme 1958). Wild vinegar flies are extremely susceptible to DDT, but by exposing a laboratory population to increasing doses generation after generation and breeding from the survivors it was possible to raise the LD_{50}—the dose that would kill half the population—within sixty generations to twenty-five times the original dose. By a series of crosses among these resistant flies, special stocks containing markers making it possible to follow individual chromosomes, and flies from the susceptible population from which the resistant flies had been selected, the twenty-seven different genotypes representing all possible combinations, susceptible and resistant, of the three pairs of major chromosomes were produced and raised in large enough numbers to test their reaction to DDT. Figure 2 shows the LD_{50} values for these combinations. Each chromosome from the resistant strain confers an increment of resistance, and two chromosomes of a pair confer twice the increment of one. But the different pairs contribute unequally. Chromosome 2 makes the largest contribution, 43 percent of the total difference between the two strains. The X chromosome makes the smallest contribution, 24 percent. Chromosome 3 accounts for the remaining 33 percent.

A chromosome-by-chromosome analysis of a polygenic system is possible only in an experimental organism where chromosomes carrying markers have been developed and where large numbers of offspring can be raised at moderate expense. But the inheritance of practically all continuously varying characters in animals bred for commercial purposes follows a pattern best explained by the polygenic hypothesis. Breeders base their programs on this assumption whether they are concerned with egg production in chickens, fat production in hogs, or milk production in cows. And in human beings, not merely pigmentation but stature, body build, and resistance or susceptibility to certain diseases all appear to have a genetic component consisting of many separate alleles, each one of small effect.

To summarize, every individual carries a set of genetic specifications—his genotype—consisting of a body of coded information that is part of the genetic system of his species. What the individual *is*—his phenotype—is the result of

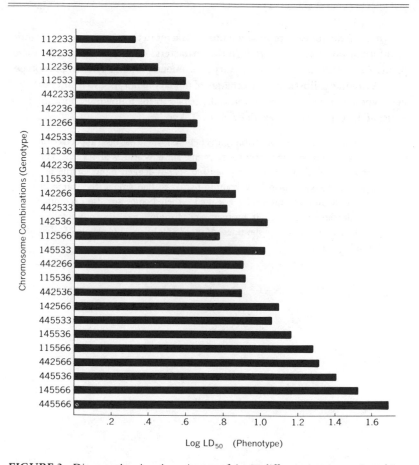

FIGURE 2. Diagram showing the resistance of the 27 different chromosomal combinations between resistant and susceptible strains of *Drosophila melanogaster* as measured by log LD_{50} in minutes of exposure to an aerosol of DDT. Three pairs of chromosomes were analyzed; 1, 2, and 3 represent chromosomes from the unselected susceptible strain; 4, 5, and 6 chromosomes from the resistant strain. Thus the genotype of a completely susceptible fly is represented by 112233 and that of a completely resistant fly by 445566. A fly heterozygous for all three pairs of chromosomes would be 142536.

the continuing interaction between his genotype and his environment from the moment of his conception to the time when we observe him. The interaction between genotype and environment is a cycle in the programmed activity of a cybernetic device—the organism. The genetic information controls the way in which the components of the organism cooperate with each other and use the environment to continue the cycle—the biotic program. The units of genetic information do not each represent a unit character in the phenotype; many units of information interact to produce the characters, each contributing its increment. As a result, most characters vary continuously around a modal phenotype.

A dramatic illustration of a change of modal phenotype as a result of altered genotype-environment interaction is described in the following quotation from a dispatch to the *New York Times* from Tokyo:

> Thanks largely to dietary habits, it is believed, there has been a spectacular change in the average Japanese physique since World War II. . . . The stereotypical diminutive, long waisted, bandy-legged figure of the old Japanese woodcut print is becoming harder to find among the young.
>
> Not only are girls longer legged—their bodies are also becoming rounder and fuller. Their brothers are changing too, becoming taller, huskier and broader of shoulder (May 17, 1970, p. 12; see also Dubos 1980, p. 16).

3

WHAT IS INHERITED?

GENETIC UNITS

In 1919 Charles B. Davenport, director of the Station for Experimental Evolution at Cold Spring Harbor, New York, published a paper in which he postulated a gene for thalassophilia, a trait that expressed itself in young men by their leaving home and going to sea. This idea of a gene for love of the sea is typical, though overly picturesque, of a way of thinking about heredity which is very ancient and dies hard. It pictures a gene as a quasi-magical entity that lies hidden in the organism and then at a predetermined moment reveals itself as a full-blown unit character. Until very recently so little was known of the nature of the gene and of its action in development that one could postulate genes having almost any sort of effect. But in the past two decades so much has been learned about just what is inherited and how it may influence development and thus produce a character, that in spinning theories of heredity one now has to keep them plausible in the light of our knowledge of genetic information and molecular biology.

To understand the mechanisms of inheritance and development, one need not be a biochemist, but one does need to know a few facts about the biochemistry involved. An organism is alive as long as it can carry out a complex series of integrated chemical processes that form part of its biotic program. These chemical reactions constitute what we call metabolism. They process materials taken in from the environment in a manner that provides the energy and the substances necessary to the functioning and repair of the organism.

Among the most critically essential molecules for life are the proteins, chains of amino acids fastened together. Amino acids are peculiar in that one part of the molecule (carboxyl group) can behave as an acid and another part (amino group) as a base. Between two such molecules a chemical reaction can take place involving the acid portion of one and the basic portion of the other. This results in their being firmly held together by what is known as a *peptide bond*. A molecule made by joining two amino acids has one free carboxyl group and one free amino group so that another amino acid may be joined to either of these sites by the formation of a second peptide bond. In this way chains, called *peptides*, may be built up. Figure 3 shows these structures diagrammatically.

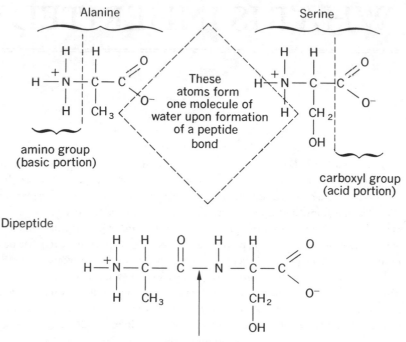

FIGURE 3. Diagram illustrating the formation of a dipeptide from the two amino acids alanine and serine. Polypeptides are formed by a continuation of this process. Another amino acid can form a peptide bond at either the carboxyl or the amino side of a dipeptide, but when polypeptides are synthesized in a living cell the first peptide bond is formed on the carboxyl side of the first amino acid, which therefore has its amino group free. The last amino acid in the chain has no peptide bond on its carboxyl side.

Chains that are very long—having a hundred or more amino acids—are usually called *polypeptides*. There are twenty different kinds of amino acids that can be strung together in any sequence into polypeptides, so that the possibilities of different permutations are without practical limits. Although polypeptides are chains of amino acids joined together, these chains are rarely extended as simple fibers. The amino acids have attractions and repulsions among themselves which cause the polypeptides to coil or twist up in very complex but specific conformations. Sometimes a single polypeptide coils up into a ball as a globular protein. More often, two or more polypeptides coil together or intertwine to form a protein. Sometimes many polypeptides align themselves together to form protein fibers. Sometimes they adhere to each other in various ways to form sheets.

The peculiar importance of proteins to the understanding of genetics is that the information that determines the sequence of amino acids in every polypeptide produced by an organism is coded in its DNA. To understand just what is inherited, it is worthwhile to give some attention to the nature of this coding and the process by which the information in the DNA is translated into protein.

As we have seen in chapter 2, DNA is a substance existing in very long filamentous strands composed of units called nucleotides. Each nucleotide is made up of three parts. One part is a special kind of sugar called deoxyribose, which contains five carbon atoms but one fewer oxygen atom than ordinary ribose. The second part of the nucleotide is a phosphate group bonded to the fifth carbon atom of the deoxyribose. The third part is a nitrogen-containing basic group bonded to the first carbon atom of the deoxyribose. In a given nucleotide this basic group will be adenine (A), cytosine (C), guanine (G), or thymine (T).

To make long strands of DNA, nucleotides are fastened together by the formation of a bond between the third carbon atom of the sugar and the phosphate group of another nucleotide. So the phosphate group holds the strand together by forming a connection between the fifth carbon atom ($5'$) of one nucleotide and the third carbon atom ($3'$) of another. Thus at one end of every strand there is a potentially free bond at the fifth carbon atom of the deoxyribose, and at the other end there is one at the third carbon. In the genetic material of all plants and animals the DNA exists in a double stranded form, the two strands spiraling around each other but oriented in opposite directions with respect to the phosphate-sugar connections—$3'$ to $5'$ along one strand and $5'$ to $3'$ along the other—so that at each end of the double helix one strand has a free bond on the fifth carbon and the other has one on the third. The phosphate groups and the sugars are on the outside of the helix; the basic groups (the sequence of which determines the information in the DNA) point inward toward each other. All along the helix each base is paired with one in the opposite strand; this pairing is specific and intimate, a guanine always pairing with a cytosine and an adenine with a thymine. Figure 4 is a diagrammatic illustration of the structure of DNA.

For every polypeptide that an organism can produce there is a section of the DNA containing a series of nucleotides the bases of which spell out the amino acids that are to go into the polypeptide and their precise sequence. It takes a series of three nucleotides to specify one amino acid. Since there are four types of nucleotides, there are sixty-four (4^3) possible triplets, more than enough to specify the twenty amino acids. All possible triplets are used. Three of them are signals for ending polypeptides. The other sixty-one are assigned to the different amino acids, some amino acids being specified by more triplets than others.

But the genetic information is not read in such a way that polypeptides are

Phosphate groups	Sugars	Bases	Bases	Sugars	Phosphate groups

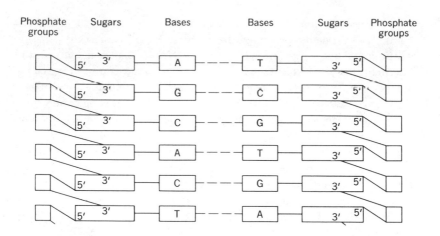

FIGURE 4. A diagrammatic representation of the structure of DNA. The three components of the nucleotides—phosphate group, sugar, and base—are indicated. On the right-hand strand the phosphate group of the top nucleotide has a bond free to attach to the 3′ carbon of the sugar in another nucleotide. In the nucleotide at the bottom, the sugar has a free bond on the 3′ carbon. On the left-hand strand the positions of these bonds are reversed. The four types of bases are adenine (A), guanine (G), cytosine (C), and thymine (T). These bases always pair specifically: A with T and G with C. Although this cross-pairing is specific, there can be any sequence of bases along the strands. The sequence in one strand determines the sequence in the other.

This diagram is two-dimensional. In the three-dimensional molecule the two strands wind around each other to form a regular double helix with ten nucleotides in each strand for each complete turn.

synthesized directly on the DNA. Instead, the sequence of nucleotides present in one functional unit of DNA containing the information coding for a polypeptide is first transcribed into another type of molecule called RNA (ribonucleic acid). Like DNA, RNA is composed of nucleotides strung together, but it differs from DNA in three important ways: (1) the sugar in its nucleotides contains one more oxygen and hence is ribose, not deoxyribose; (2) RNA does not exist in long, stable double helixes but is usually single stranded; and (3) although there are four different bases, there is no thymine in RNA; it is replaced by another base, uracil (U). When the message in the DNA coding for a given polypeptide is transcribed into RNA, it comes off as a single strand having a sequence of bases complementary to those in the strand of DNA on which it was made. If the DNA has the sequence GCAAGT, the corresponding messenger RNA (m-RNA) will have the sequence CGUUCA.

In order to translate m-RNA into a polypeptide, it has to be processed by the protein-synthesizing machinery of the cell. This machinery is complicated and delicately controlled, but the major elements and the way in which they interact to carry on the process can be described in understandable, if oversimplified, form. The machinery consists of an organelle, a set of specialized molecules, and a group of enzymes. The organelle, which is called the *ribosome*, is a tiny yo-yo-like object composed of two hemispheres of slightly different size. The m-RNA molecule seems to be threaded in the crevice between the two hemispheres. As the m-RNA moves along this crevice, the specialized molecules—called transfer RNA, or t-RNA—move up one by one, each bringing the specific amino acid called for by the triplet of m-RNA nucleotides which has just entered the crevice of the ribosome. While an amino acid is in position, it is linked by a peptide bond with the amino acid that entered just before. The process continues until the entire series of triplets of the m-RNA has been translated into a polypeptide chain. Figure 5 illustrates this process diagrammatically. When the last amino acid has been bonded to the chain, the chain is released from the ribosome and then usually coils up either by itself or with other polypeptides of the same composition or of other composition made from other messengers, and a protein molecule results.

There are many classes of proteins. One of the most important comprises the enzymes. These are catalysts that determine the chemical reactions that take place in the organism and control their direction, speed, and interrelations. The enzymes make metabolic activity possible and determine its nature and efficiency. Another class consists of the structural proteins. These serve as building materials and are found in the microtubules that act within the cell as both skeleton and muscle, in cell membranes, in connective tissue, in cartilage, tendons, hair, and nails. Structural proteins are often modified after synthesis by having nonprotein molecules fastened to them as a result of enzymatic action, but their amino acid sequences are constructed by the protein-synthesizing machinery of the cells. A third class of proteins includes those that act as regulatory messengers, such as hormones and receptors.

The mechanisms for transcribing the information in DNA into m-RNA and its translation into protein were worked out in bacteria in the 1960s (Watson 1976). In these simple organisms the DNA is not segregated in a nucleus. While it is being transcribed, m-RNA is threaded on the ribosome and translation to the polypeptide gets under way before transcription is completed. The m-RNA in bacteria is very short lived, being degraded soon after the translation of the prescribed number of copies.

To illustrate the process of gene regulation, let us look at the gene in *E. coli* which codes for the enzyme beta-galactosidase. This enzyme is able to split the disaccharide lactose into its two component simple sugars, galactose and glucose. Unless this splitting takes place, the cell cannot utilize lactose either as a source of energy or as a raw material for synthesizing other molecules. If *E. coli*

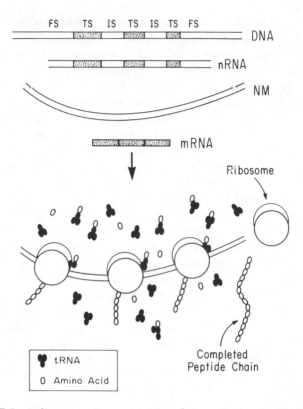

FIGURE 5. A diagrammatic representation of the mechanisms by which information coded in the DNA is transcribed into n-RNA, processed into m-RNA and then translated into protein. At the top of the figure is a section of DNA consisting of three translated sequences (TS) interrupted by two intervening sequences (IS) and flanking sequences (FS) on either side. This whole section is transcribed into n-RNA, the precursor of m-RNA. Before the latter is transferred through the nuclear membrane (NM), the portions corresponding to the flanking and intervening sequences are processed out. Beginning at one end of an m-RNA molecule, ribosomes progress along it independently—or the m-RNA moves through one ribosome after another. As the movement occurs, t-RNA molecules, each bearing a single amino acid, come to the point of contact of the ribosome and the m-RNA and deliver their amino acids one by one to the growing polypeptide chain. As several ribosomes move along one m-RNA molecule, each has dangling from it the portion of the polypeptide chain completed up to that point. When a ribosome arrives at the end of the RNA molecule, it leaves the molecule and releases the completed polypeptide. The sequence of bases in the m-RNA molecule determines just which of the twenty different amino acids will be added next to the growing peptide chain. The m-RNA molecule threaded through the ribosomes in the diagram is on a different scale from the molecule shown immediately below the nuclear membrane.

cells are growing in medium containing no lactose, the gene for beta-galacto-sidase will be inactive because a repressor protein molecule, acting as a regulatory messenger, sits on the DNA at the beginning of the gene and prevents the polymerase enzyme from transcribing it. If lactose appears in the medium, some molecules will get into the cell. The repressor molecule has a stronger affinity for the lactose molecule than it does for the site on the DNA, and when a lactose molecule comes in contact with one, it lets go of the DNA and allows the polymerase enzyme to transcribe m-RNA molecules. These are then translated into beta-galactosidase, which splits lactose molecules and makes the simple sugars available to other enzymes in the cell.

If a limited supply of lactose is exhausted, some of the repressor molecules will lose their holds on lactose and regain their affinity for the repressor site at the start of the gene. This will shut off transcription and no more beta-galacto-sidase will be made unless a new supply of lactose appears in the medium.

Beta-galactosidase is a catabolic enzyme. It takes large molecules apart into simpler ones. Other genes code for anabolic enzymes that synthesize more complex molecules by putting simpler ones together. The regulation of anabolic genes is somewhat different from that of catabolic ones. There is a group of enzymes that cooperate to synthesize histidine, one of the twenty amino acids necessary for the translation of m-RNA into a polypeptide. In a growing cell these enzymes will be needed and the genes that code for them will be available for transcription by RNA polymerase. There will be no repressor molecule sitting on their repressor site. Floating around in the cell are inactive, but potential, repressor molecules that become active when combined with a histidine molecule. If the concentration of histidine reaches a critical level, either as a result of oversynthesis, underutilization, or the diffusion of histidine into the cell from the surrounding medium, some of the inactive repressors will team up with idle histidines, form active repressors, and one of these will occupy the repressor site and shut off the transcription of the genes coding for the histidine-synthesizing enzymes.

Thus in bacteria enzymes that synthesize are transcribed and translated when the products they synthesize are not present in excess, and their transcription and translation are shut off when there is no need for them. Enzymes that tear down substances are transcribed and translated only when the substance is present to be worked on. The system is simple and thrifty. Genes are turned on when their products are needed and off when those products would simply accumulate and get in the way. It uses the materials available with a minimum expenditure of energy.

For many years attempts were made to explain gene action in eukaryotes— the higher organisms including man—by applying the mechanisms of gene regulation known in bacteria—the prokaryotes. These were not successful. In the cells of higher organisms the DNA is confined in a nucleus set off by a membrane. All transcription of DNA takes place in the nucleus; all translation of m-RNA occurs outside the nucleus in the cytoplasm where the ribosomes are found. But in the nucleus there is no m-RNA, only a snarl of excessively long strands of RNA that long defied classification or recognition. If messages are communicated from the cytoplasm to the DNA, they must in some fashion get across the nuclear membrane. In addition, in higher organisms messages must be transferred from cell to cell and this means getting them across two *cell* membranes. All of this is hard to explain on the simple bacterial model.

Furthermore, there are two profound differences between the DNA of prokaryotes and that of eukaryotes. The first of these pertains to the way the molecule is arranged. In bacteria the DNA molecule is a long thread; its ends are joined to make a continuous circle, folded back and forth into a compact mass in one portion of the cell. In eukaryotes the DNA is not a continuous circle but is broken up into several separate, two-ended pieces of varying length—the chromosomes. Each of these pieces is organized into a series of repeating units known as nucleosomes. A nucleosome consists of a small spherical core of a special kind of protein called histone around which a double loop of approximately 200 base pairs of DNA is wound. Between nucleosomes there is a length of about 50 base pairs connecting them, giving a structure resembling a string of beads (Felsenfeld 1978). In humans the DNA in each cell is divided into 46 pieces—chromosomes—and within each there are hundreds of thousands of nucleosomes. The histones in each nucleosome are like those in every other, but associated with this long necklace are other proteins, known appropriately as nonhistones, which form a matrix around it. These proteins are highly heterogeneous and in constant flux, different components being synthesized and degraded, added and subtracted in continuous turnover (Stein, Spelsberg, and Kleinsmith 1974). This whole complex string can be extended as a chain of nucleosomes or twisted into coils, supercoiled on itself and then folded back and forth to become a tightly condensed mass. The chromosomes are at their greatest condensation at the time of cell division; in a growing or active cell they tend to be uncoiled and extended.

The second difference has to do with the sequencing. If DNA in solution is broken into pieces a few hundred base pairs long and heated, the two strands come apart. If the mixture is allowed to cool, the single strands will pair with their appropriate partners. If this is done with bacterial DNA, the re-pairing or annealing, measured on a logarithmic scale, proceeds at a constant rate, indicating that every single strand has an equal chance of finding a partner. If the same procedure is followed with human DNA—or DNA from any of the higher organisms—the rate of annealing is at first rapid, slows down and then speeds up again, proceeding in several such steps to completion. This behavior means that some sequences are repeated—some of them many, many times—have, therefore, many potential partners, and hence pair rapidly. Others are repeated fewer times; still others are unique to a given genetic complement and are the last to pair (Schmid and Deininger 1975, Britten and Kohne 1968). The repeated and unique sequences are scattered throughout the DNA. Close to half of human DNA can be classified as repetitive.

All these facts became known in the 1970s and were extremely puzzling. Two astonishing discoveries made in 1977 have contributed greatly to clearing up the confusion. First, it is now known that in eukaryotes the DNA coding for almost all polypeptides is not continuous but is interrupted from one to several times by stretches of DNA that are never translated. This contradicts an idea that had been part of the credo of molecular genetics for more than two decades. The notion of the gene as an intact sequence of nucleotides defining the order of amino acids in one polypeptide was beyond question. This had been shown to be the case in bacteria. It was assumed to be universal. The discovery of *intervening sequences* was a rude shock. Second, when a segment of DNA is transcribed in the nucleus, not only are these intervening sequences transcribed along with the gene in which they are embedded, but the transcription also includes *flanking sequences* at each end of the gene. This results in an extremely long molecule of RNA, often several times the length of the sequences necessary for translating the polypeptide. Within the nucleus this long molecule is trimmed, cut, and spliced in one or more steps to produce the m-RNA ready for translation. The last step in this process is the transfer of the m-RNA through the nuclear membrane into the cytoplasm. This second discovery, although it immediately suggested why m-RNA could not be found in the nucleus, made necessary an equally painful revision of the doctrines of molecular genetics for it had long been assumed that m-RNA was a mere transcription of the information in the gene. Suddenly, one had to make room for levels of processing and discrimination between gene and messenger. The summaries and comments on these discoveries by Leder (1978) and Crick (1979) give an idea of the shock that ran through the molecular genetic community when it was confronted with these new facts.

By no means all the details of the structure of eukaryotic DNA, of the control of its transcription or the processing of the resulting RNA are yet known or understood. But what we do know now makes it possible to describe the probable structure of human genes and the mechanism of transcription in a way that suggests an understandable system of gene regulation.

To get a picture of this more complex gene regulation in higher animals, some work done on the chicken is easier to explain and understand than anything known directly in humans, although there is no question that the mechanisms are the same in both birds and mammals. In the oviduct of the hen a process takes place which provides a classic illustration of the transmission of information from component to component to further the functioning of a cybernetic process. The ovary delivers egg cells to the oviduct where, along with their growth into fully formed yolks, they are also provided with large quantities of the protein ovalbumen. This constitutes a substantial portion of the white that surrounds the yolk and it is synthesized by cells in the walls of the oviduct. As the ovary puts egg cells into the assembly in the oviduct, it also secretes into the hen's circulation molecules of progesterone, a steroid hormone, a fairly small molecule, precisely the same hormone molecule that is active in human females in maintaining pregnancy. In the hen these molecules diffuse from the blood into the cells of the oviduct wall. In solution in the cytoplasm of these cells are protein receptor molecules, synthesized within the cells themselves, each composed of two dissimilar parts. Each of these two parts has a site with an affinity for progesterone. In the absence of progesterone these receptors are inactive, but as progesterone becomes available, each of the sites takes up one molecule of progesterone and this four-way combination then becomes able to cross the nuclear membrane and deliver a message. Once in the nucleus, one part of the protein receptor has an affinity for the nonhistone protein at the site on the chromosome where the gene coding for ovalbumen begins; the other reacts with the DNA at that site and makes it available for RNA polymerase. The long RNA molecule, including flanking and intervening sequences, is transcribed. This is then processed to produce m-RNA that is put through the nuclear membrane, translated on the ribosomes in the cytoplasm; and this ovalbumen is then secreted into the oviduct to become part of the coating of a yolk to produce a complete egg ready to be provided with a shell. When the ovary stops putting egg cells into the oviduct, it stops secreting progesterone. The hormone-receptor packages gradually fall apart and are not replaced, and the synthesis and secretion of ovalbumen slows down and eventually ceases (O'Mally and Schrader 1976).

That this type of gene regulation exists in humans is demonstrated by a rare hereditary disease termed testicular feminization. In this disorder individuals with a male chromosome complement develop as females so far as outward phenotype is concerned although they are sterile. It has been shown that this occurs because in these individuals those cells, which in the normal male respond to the male hormone testosterone, completely lack the protein receptors that must combine with testosterone to enter the nucleus and initiate transcription of m-RNAs capable of producing the morphology of the male phenotype (Wilson and MacDonald 1978, Griffin 1979).

Hormones in the form of small molecules such as steroids that can diffuse into the cell may react with receptors found inside. Larger hormone molecules such as proteins cannot get through the cell membrane by themselves and are aided in delivering their messages by receptors in the cell membrane which bind with them and either bring them bodily into the cell or set off some sort of reaction within the cell which conveys the appropriate message.

There are many kinds of messenger-receptor interactions but few of them are completely understood. One that we know reasonably well and which illustrates the complications of gene regulation concerns the control of the concentration of cholesterol in the human body. Cholesterol is a fatty molecule, insoluble in water, which is a necessary constituent of all cell membranes. Every time a cell divides it needs enough cholesterol to allow for the doubling of the membrane area. If the amount in circulation in the body becomes too great, harmful deposits of it turn up in such forms as gallstones or as plaques on arterial walls that lead to atherosclerosis and heart attacks. Even an undividing cell needs cholesterol for maintaining its membrane in good repair and most cells are able to produce the enzymes that synthesize it. This is one source of cholesterol. The other is from the food we eat, through absorption by the small intestine. Cholesterol entering the blood plasma from the intestine is taken up by a special carrier molecule called low density lipoprotein and this combination circulates in the blood. Most active cells have receptors composed of protein made within the cell and built into the cell membrane in such a way that they are partially exposed on its surface. Lipoprotein-cholesterol molecules circulating in the blood come in contact with these receptors and bind with them. Groups of receptors, acting together, then enter the cell bringing the bound molecules with them. Inside, the cholesterol is removed from the package and made available to the cell. As long as the concentration of available cholesterol inside the cell is at a satisfactory level, it sends two inhibitory messages into the nucleus. One shuts off the production of the enzymes that can synthesize new cholesterol; the other slows down the production of new receptors for the cell membrane. When the need for cholesterol within the cell rises, these inhibitions are relaxed. More receptors are produced and more lipoprotein molecules carrying cholesterol are brought in from the plasma. If the plasma cannot furnish the cell with sufficient cholesterol, the genes coding for the synthesizing enzymes will be transcribed, the resulting RNA processed and exported from the nucleus and this m-RNA will be translated on the ribosomes. The end result is that, normally, the level of cholesterol is kept in dynamic equilibrium with its two sources by a system of feedbacks. Cells are provided with a constant and sufficient supply; the accumulation of excessive quantities is avoided (Brown and Goldstein 1976).

This complicated system has been worked out through the study of individuals who have inherited an allele producing defective receptors and who, therefore, neither remove cholesterol from their plasma nor shut off its intracellular synthesis. Heterozygotes with one such allele have abnormally high levels of cholesterol in their plasma and suffer from cholesterol deposits in skin, arteries, and the cornea of the eye. Homozygotes show these abnormalities in exaggerated form and have poor prognosis and greatly reduced life expectancy (Fredrickson, Goldstein, and Brown 1978).

In neither case—that of the steroid hormones nor that of cholesterol stabilization—do we have a complete picture of all the details of gene action. But it is now possible, with the aid of the new discoveries in DNA sequences and of RNA processing, to build a reasonable hypothesis for gene action in eukaryotes. The intervening and flanking sequences show us that there is much more information in the genetic complement than merely the definition of amino acid sequences in polypeptides. That this extra information is one of the distinguishing features between prokaryotes and eukaryotes strongly suggests that it is needed for the more complex functioning of the nucleated cell and the multicelled organism. It is almost certain that the extra information accounts for the repeated sequences and that the unique sequences are those that code for the ordering of the amino acids in the polypeptides.

To replicate DNA, which must be done every time a cell divides—many billions of times to produce a human phenotype—requires material and energy. In the biological economy it is expensive. If the repeated sequences were not necessary, they would have been weeded out long ago. E. *coli*, with only about one eighteen-hundredth the amount of information present in the human genetic complement, is able to synthesize all twenty of the amino acids from simpler substances. Humans can synthesize only nine of these and depend on their diet for the eleven others. Sometime in the history of life we lost the DNA sequences coding for the necessary synthesizing enzymes. Rats can synthesize ascorbic acid; humans cannot and depend on their diet to provide it. At some later time we lost the recipe for the enzyme necessary to make it. But we keep our repeated sequences. This means that in the long run it has been more advantageous to the individual to retain the repetitive sequences than it has been to hold on to the genes capable of synthesizing some substance constantly available in the diet anyway.

What do these repeated sequences do? We do not know. It has actually been suggested—and the idea has been seriously entertained by so awesome a personage in this field as Francis Crick—that these repeated sequences serve no important function and are merely "junk" (Orgel and Crick 1980). This seems highly unlikely. They are scattered throughout the genetic complement between and within the genes and it seems much more likely that they constitute the machinery for gene control—a sort of filing and retrieval system for "conventional genes," those DNA sequences that are ultimately translated into polypeptides. It is probably a misnomer to call everything but the old-fashioned "genes" repeated sequences. Some of these are repeated millions of times, others hundreds of thousands, perhaps others so few times that they approach the unique sequences in frequency. Probably it is preferable to think of translated and untranslated sequences. These untranslated sequences may act as tabs or labels showing where the sequences to be translated are located. They may determine what nonhistone proteins sit at which sites along the chromosome and when during development and metabolic activity these proteins are to be removed or changed. This would allow hormones and other messengers to know where to attach to activate a given gene. When a gene is activated and transcribed, the untranslated sequences that are processed out in forming the finished m-RNA can carry other messages specifying under just what circumstances that m-RNA is to be released for translation. In the translated sequences the triplets act as words specifying amino acids. The repeated sequences may very well be longer words in a different code which coordinate the actions of different genes within the same and in different cells to orchestrate the development and the functioning of the organism. In this code some words are used over and over. These are the highly repetitive sequences. Others occur less often; these are those of intermediate repetition. There may be others that are needed so rarely that we do not recognize them as repeated.

However the code for gene regulation works, in higher animals it must be able not merely to turn genes on and off as occurs in bacteria; it must also be able to inactivate permanently whole blocks of genes in different types of cells. Hemoglobin is synthesized only in young red cells; in all other cells of the body the hemoglobin genes are never activated. Neurons never produce insulin. Lymphocytes do not produce melanin. This selective, permanent inactivation of the greater part of the genetic complement in the somatic cells is part of the process of differentiation that occurs during development. We know very little about the exact mechanisms used, but it is clear that they must operate with great precision and, to accomplish this, much very precise information must be coded in the genetic complement. This must be found in the untranslated sequences.

Among the untranslated sequences there are some units whose functions we do know and understand. One such type of unit codes for the t-RNA molecules that bring the single amino acids to the proper position on the ribosome during protein synthesis. Another type of RNA that is not translated is ribosomal RNA (r-RNA), which makes up a substantial proportion of both hemispheres of the ribosome. Finally, although it has not been conclusively demonstrated, it is very probable that some RNA molecules carry out other regulatory functions.

So this digression into information theory and molecular biology enables us to see that what is inherited are functional units of information coded in the DNA. These can be classified into six groups corresponding to (1) enzymes, (2) structural proteins, (3) polypeptide hormones, (4) ribosomal RNA, (5) transfer RNA, and (6) untranslated regulatory messages. These units of genetic information, beginning their interactions within the cytoplasm of the egg and continuing it through the process of development in a specific environment produce the organism with its unique phenotype. All other substances synthesized by the organism are the result of the activities of these six types of genetic messages. These are what we inherit. We do not inherit kinky hair or thalassophilia. What we do inherit are functional units of information, which, interacting with each other and with the environment, determine the probability of our hair's being kinky and just might conceivably influence the likelihood of our running away to join the Navy.

DEVELOPMENT: CONSTRUCTING THE PHENOTYPE

Development is the process by which the information in the genetic complement interacts with the information in its environment to form an individual. It begins with a series of about a dozen successive, largely synchronous divisions of the fertilized egg resulting in a mass of cells numbering somewhere between 5,000 and 10,000. The exact topology of this mass varies among different classes of vertebrates from a hollow sphere to an elliptical mat, the variation being largely an adaptation to the amount and nature of the nutritive material in the egg. Whatever the shape, soon after synchronous divisions stop, a set of movements among the cells rearranges them so that three groups or layers become recognizable. These are the famous trinity of embryology—the ectoderm, the mesoderm, and the endoderm. The first gives rise to the skin and the nervous system; the second develops into skeleton and muscle; and the third produces the gut and its accessories.

The most important aspect of this rearrangement and the sorting out of the three layers—the process known as *gastrulation*—is that it results in what is known as *primary induction*. The mesoderm lies under the ectoderm, and the two interact to produce the primary organization of the embryo. The ectoderm develops into the beginning of the nervous system with a head-tail orientation, a brain at the head end which gradually differentiates into a fore, mid, and hind portion and a spinal cord extending behind it. The mesoderm under this rudimentary nervous system shows at the same time the beginning of a notochord; on either side of it the somites, segments that later become the vertebrae of the spinal column, begin to develop.

What is most significant about primary induction from our point of view is that the mesoderm and the ectoderm interact mutually to produce it. Neither will show its own organized development without the other. In amphibian embryos, where it is possible to do experimental operations, an extra piece of mesoderm from another embryo implanted under a part of the ectoderm which is normally not in contact with mesoderm will induce a second embryo, and the two may go on developing to the hatching stage, appearing as Siamese twins attached at the belly. Obviously the type of interaction that takes place between the groups of cells in induction must be mediated chemically. But so far, in spite of much painstaking work, it has not been possible to isolate and identify an inducing substance. It is clear, however, that the process must be under genetic control. Different groups of animals carry out the process in their own distinct ways, and mutations are known which cause the process to go in an abnormal direction. In mice, for example, there is a mutant allele (Brachyury, T) that is dominant in heterozygotes, in which it produces a short tail. But when homozygous it is lethal, and homozygous embryos show abnormalities soon after gastrulation; the notochord fails to develop according to the' normal pattern (Hadorn 1961, p. 166).

This primary inductive process, which organizes the fundamental embryonic architecture at gastrulation, is only the first of a complex series of inductions that succeed each other in hierarchical fashion throughout embryonic development. In vertebrate embryos generally, soon after the forebrain becomes distinct, two processes grow out of it, one on either side, at the ends of which are the optic cups. The inside of this optic cup later becomes the retina of the eye. The optic cup lies close to the skin, and the skin immediately over it becomes the lens of the eye. In the absence of an optic cup, no lens will form, and an optic cup transplanted to another part of the embryo will induce a lens to form in whatever skin lies immediately over it. Another instance of an inductive relationship has been analyzed in mutant mice. Another mutant that affects the tail—Danforth's short tail (Sd)—is lethal when homozygous, but mice heterozygous for the allele not only have a short tail but also some abnormalities of the urinary system (Hadorn 1961, p. 128). These abnormalities go all the way from a slight reduction in the size of one kidney to complete absence of both kidneys. In the latter extreme case, of course, the gene proves to be lethal even in heterozygous condition. What is most interesting is that in the heterozygotes, whenever even a small kidney has formed, it is connected with the bladder by the tube known as the ureter. During development the ureter grows toward the tissue that becomes the kidney. Whenever no kidney has formed, the ureter has not grown as far as the kidney site. The growing ureter appears to induce the formation of the kidney in the cells at the presumptive kidney site. Without the inducing effect of the ureter no kidney forms, although the tissue at the kidney site is quite competent to form a normal kidney under the proper inductive stimulus. This case is also of great interest because it illustrates that gene action is not an all or nothing matter. Sometimes the induction fails completely; other times it occurs normally; all sorts of intermediate patterns are also observed.

Sometimes induction is normal on one side of an embryo and completely abnormal on the other. In a given individual, abnormality is a probability, not a certainty.

The complex succession of inductions which takes place in embryonic development is exceedingly difficult to follow in all its details, but there are many other observations that persuade us that this is what goes on. Phenocopies (see chapter 2) are often the result of the blocking or abnormal direction of an induction. The effect of Thalidomide on the human embryo is confined to the period from the thirty-fifth to the fiftieth day of gestation, and the type of effect changes from the beginning to the end of the period. On the thirty-fifth day the effect is most likely to be an abnormality of the ear; a few days later it will be a shortening or suppression of the long bones of the arm. In a few more days the effect shifts from the arms to the legs, and at the end of the period the abnormality is likely to be confined to the thumb. In some way the chemical structure of the Thalidomide molecule gives a signal that interferes with the normal development of the induced beginnings of these organs, and the differences in effects through time indicate the periods during which the different induced rudiments are most susceptible to influence. Many other illustrations of chemical or physical interference with normal development could be cited in many different species of animals. All of them are explicable by postulating that the organism is an exquisitely designed cybernetic system in which the cycle is kept on the normal path by the transmission and receipt of information between the components of the system. Most of the messages are chemical, and the introduction of a wrong message at a critical time can disrupt the normal program to a greater or lesser extent, all the way from causing slight abnormality to producing death.

The foregoing discussion clearly illustrates the statement that a phenotype—an individual—is the result of the interaction between the genotype and the environment. The information in the DNA of the genotype is programmed to produce the phenotype of the species when the development proceeds in the usual environment of the species, but the program may be badly distorted by unusual substances in the environment, of which Thalidomide in the bloodstream of a mother carrying an embryo is an example. Another example is afforded by certain amphibian embryos. If they are placed in water containing lithium above a certain concentration, the notochord fails to develop in much the same way that it fails in mouse embryos homozygous for the short tail T gene mentioned earlier. However an embryo develops, whether it lives or dies, whether its phenotype is normal or abnormal, at any given stage its phenotype is the result of the interactions that have gone on between genotype and environment up to that time. Throughout development an embryo is a unit and develops as a functioning whole. Each successive stage is the result of changes in the preceding stage, and these changes must occur in a pattern that retains the integration of the whole. The process is continuously dynamic and integrated. It is not an aggregation of several independent processes, each concentrating on the production of some unit character. With the end of embryonic life—birth in mammals or hatching in many other kinds of animals—the process of development seems to become less radical and rapid, but the same processes go on even though in somewhat less dramatic fashion. Infancy, childhood, maturity, and senescence—each represents a stage that grows out of the one before, and the interaction between genotype and environment continues, at every moment under the strong influence of all that has occurred up to then.

THE MOLECULAR GENETICS OF POLYGENES

Let us see to what extent we can picture a polygenic system in terms of hereditary units and the interactions between them and between the organism and the environment. The epidermis—the outer layer of human skin—is made up of an upper horny layer (*stratum corneum*) composed of the remains of dead cells that were formerly part of the live layer beneath (*stratum Malpighium*). Among the live cells of the under layer are melanocytes that produce melanin, a dark brown pigment. The starting point for the synthesis of melanin is the aromatic amino acid tyrosine, which is oxidized by the enzyme tyrosinase to dihydroxyphenylalanine (dopa). Dopa then undergoes further alteration and winds up as indole-5-6-quinone. Molecules of the latter polymerize to form melanin. Within the melanocytes small packages termed melanosomes form, and then melanin is synthesized within them. The final pigment granule is a combination of melanin and protein. The granules can differ in size and number and in the composition of the protein portion and the degree of polymerization of the melanin molecules. The ratio of melanocytes to epidermal cells may vary. Melanin production in the melanocytes is stimulated by certain hormones. If the outer horny layer of the epidermis is sufficiently unpigmented to permit the transmission of ultraviolet light, the effect of such radiation on the live cells beneath is to stimulate the melanocytes to synthesize more melanin. Furthermore, an important constituent of the horny layer is keratin, a complex protein produced by the epidermal cells. The amount and type of keratin in the horny layer can influence its transparency to ultraviolet light. Thus the action of the epidermal cells in producing keratin can determine the degree to which the skin will respond to ultraviolet irradiation by producing more melanin. In addition to occurring in the epidermis, melanocytes are found also in the retina of the eye. The mutant allele that causes albinism turns off the production of melanin in both the skin and the eye. Another mutant, for ocular albinism, suppresses melanin production in the retina but not in the skin. The first is an autosomal gene, the latter sex-linked, showing that two separate control mechanisms are involved.

A moment's pondering of the information in the preceding paragraph makes it clear that there must be many different functional units in the DNA which interact to determine the level of pigmentation in any one individual. It is impossible to say just how many enzymes, structural proteins, and regulatory molecules take part in the system. But it would not be unreasonable to postulate two or three dozen. So to picture a polygenic system controlling pigmentation one does not have to resort to a simple-minded model of thirty or forty pigmentation genes each contributing one unit of darkness. One can see that differences in the effectiveness of many interacting enzymes, structural proteins, and regulatory molecules can be combined in different ways to produce a graded series of genotypes and corresponding phenotypes.

Nor does one have to think of merely a plus and a minus allele for each one of the functional units of the system. There is good evidence for concluding that every message coded in the DNA exists, in any sizable population, in numerous versions, forming a spectrum grading from grossly defective alleles—such as the one for albinism—at one end, through the slightly deviant, to the normal at the other end. And the normal is probably not a single version of the message but a collection of slightly different alleles, each giving the usual phenotype but showing subtle differences in efficiency depending on the alleles with which they are associated and the environment with which they interact.

Until recently the only direct evidence for differences between alleles at the molecular level consisted in changes in amino acids of a protein assumed to be the result of changes in the nucleotides of the corresponding DNA. Over the past two decades many examples of such amino acid substitutions have accumulated. Among human proteins the best known from this point of view are the hemoglobins. In the normal adult more than 95% of the pigment in the red cells of the blood consists of hemoglobin A, a complex molecule composed of four polypeptide chains called globins tangled together: two alpha chains and two beta chains that differ from each other slightly in length and somewhat more in amino acid sequence, coded at separate sites in the DNA. Associated with each chain is a small nonprotein molecule, a heme, containing an iron atom. The whole assembly is able to take up and release oxygen and transports it from the lungs to other tissues.

The beta globin chain consists of 146 amino acids and its normal sequence has been established by analysis of beta chains from a substantial sample of individuals. In addition, well over one hundred different abnormal beta chains have been identified and analyzed. The great majority of these differ from the normal in having one single amino acid at a given position in the sequence replaced by another. In almost all these cases it has been definitely established exactly which amino acid has been replaced by which other.

Many of these beta chain mutations produce pathological symptoms in the individuals who carry them. They have a tendency to produce anemia, because the instability of the molecules containing the mutant chains makes the red cells susceptible to premature fragmentation. Most of these deleterious forms of hemoglobin were found as a result of analyzing the hemoglobin of patients suffering from anemia. In a majority of these cases the amino acid substitution is such that it changes the electric charge on the hemoglobin molecule. This makes it possible to distinguish the mutant from normal hemoglobin by electrophoresis. Several screening programs have been carried out in which hemoglobins of asymptomatic individuals selected at random have been subjected to electrophoresis. Some beta chain variants found by this method are substantially normal in that their carriers suffer no ill effects, although most of these variants have, in addition to their electrophoretic behavior some physico-chemical characteristics slightly different from the normal. Nevertheless, there is reason to believe that the beta chain mutants that we know are not an unbiased sample of the variants existing in the population. Most of the known variants have a different charge from the normal chain or produce an unstable molecule. We have no practical method of looking for variant hemoglobin chains in which one amino acid has been replaced by a very similar one. Such a substitution would only rarely result in a difference in charge and would be likely to have little effect on the structure of the molecule. A search for such variants would entail the complete analysis down to the last detail of the amino acid sequence in hemoglobins from many hundreds of people selected at random. This would be an extremely costly project.

Two of the most common beta chain mutants are hemoglobins (Hb) S and C. Both contain substitutions for the glutamic acid in the sixth position in the beta chain. In Hb S the substituted acid is valine. This reduces the number of negative charges on the chain by one, for glutamic acid is negative and valine neutral. In Hb C the substitute is lysine, an acid that carries a positive charge. This gives the C chain two fewer negative charges than the normal beta chain. It is possible to explain these two substitutions as the result of single nucleotide substitutions in the DNA. The m-RNA triplet GAG codes for glutamic acid. If the A is changed to U, the resulting GUG will code for valine. If the first G of GAG is changed to A, the resulting AAG will code for lysine. What is extremely interesting is that if the final G of GAG were changed to either C or U, the result would be to substitute aspartic acid for glutamic acid. This substitution would not change the overall charge of the chain and would be less likely to modify the tertiary structure of the molecule than any other substitu-

classical genetics. Some very early cases showed the possibility of three or more alleles at a single locus; for example, the ABO locus for human blood groups and the locus in *Drosophila* producing a graded series of pale eyes going all the way to white. Wild-type or normal alleles were originally assumed to be all alike, but as early as 1943 some very careful experiments with *Drosophila* by Curt Stern and E. W. Schaeffer showed that this was not true. They coined the term *isoalleles* to describe wild-type alleles that appeared superficially to be identical but which could be shown to act somewhat differently under carefully controlled conditions. Stern and Schaeffer had no idea how their isoalleles differed chemically. Later, it became clear that many of the asymptomatic hemoglobin variants, the active revertants of tryptophan synthetase in *E. coli* and the A and B forms of human G6PD were precisely what Stern and Schaeffer postulated—alleles that are substantially normal but which have slight, subtle differences between them. But the more recent discoveries of intervening sequences and of the nuclear processing of RNA make it possible now to understand how alleles may differ in their effects on the phenotype and yet show no differences in the sequences of amino acids in the polypeptides for which they code. Stern and Schaeffer's hypothesis of an array of isoalleles is fast becoming a reality of molecular biology.

To illustrate this new insight let us look at another type of abnormal hemoglobin allele. For many years it has been known that in populations around the Mediterranean there is a special kind of anemia called thalassemia. As techniques of molecular genetics developed, it became possible to show that in this disease there were no abnormal hemoglobin chains. The abnormality lay in the ratio of alpha to beta chains. In most cases from the Mediterranean area the proportion of beta chains was drastically reduced, sometimes to complete absence, and the name beta thalassemia was given to the disorder. A somewhat similar reduction of alpha chains, rare in the Mediterranean but more frequent in Southeast Asia, is also known, but for our purposes, let us concentrate on the beta variety (Bank, Mears, and Ramirez 1980).

There are still many things about beta thalassemia that we do not understand but progress in unraveling its complications has recently been very rapid and new revelations will undoubtedly be made in the near future. The techniques of DNA analysis with restriction enzymes and the ease with which DNA can now be sequenced have led to the knowledge on which the diagram in figure 6 is based (Van der Ploeg et al. 1980). In addition to the alpha and beta chains, there are three others: G-gamma, A-gamma, and delta. The hemoglobin formed with two alpha and two beta chains, hemoglobin A, constitutes about 97 percent of the hemoglobin found in the red cells of normal adults. About 2 percent of the hemoglobin in adults, hemoglobin A_2, is formed with two alpha chains and two delta chains. The remainder, about 1 percent, consists

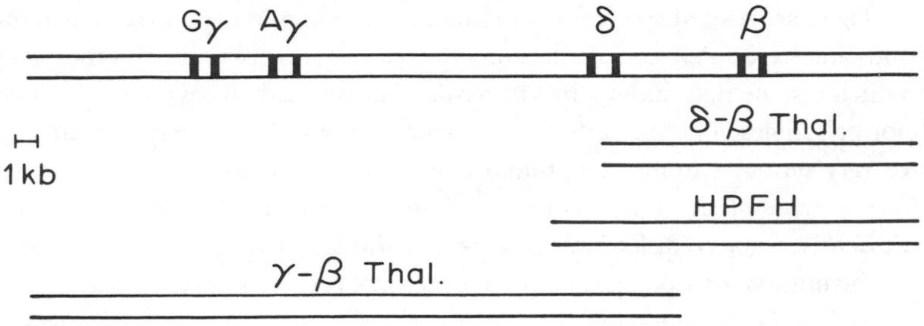

FIGURE 6. Diagrammatic representation of the DNA of the gamma-delta-beta hemoglobin genes on human chromosome 11. At the top are the intact, normal genes: G-gamma, A-gamma, delta, and beta. The filled-in areas represent the translated sequences; the open areas untranslated sequences, an intervening sequence within each gene, and flanking sequences on either side. Below are indicated those sections of the normal sequences which have been deleted in delta-beta thalassemia, hereditary persistence of fetal hemoglobin, and gamma-beta thalassemia. 1 kb equals 1,000 base pairs. (Based on data from Bank, Mears and Ramirez 1980, *Science* 207:486, and Van der Ploeg et al. 1980, *Nature* 238:637.)

Since the papers on which this diagram was based were published in February 1980, it has been shown that the globin genes figured here have two intervening sequences, not one (Proudfoot et al. 1980). This indicates how rapidly knowledge in this field is advancing. The change does not, however, vitiate the principle illustrated by the diagram.

of molecules of two alpha and two gamma chains—about half G-gamma and half A-gamma—known as hemoglobin F. The F stands for fetal as this is the type of hemoglobin in the fetus during the latter part of gestation.

The top line of the diagram represents a linear extension of about 40,000 base pairs of DNA in human chromosome 11. Within it are the genes for the two gamma chains, the delta chain and the beta chain. Each gene contains about 450 base pairs coding for amino acids and in each case this sequence is interrupted by an intervening sequence of about 1,000 base pairs. In case you are wondering, the alpha chain genes, of which there are two, are at a completely different site on chromosome 16. In the normal individual by the third month of gestation the alpha genes and the gamma genes are producing equal numbers of chains to form hemoglobin F. At the time of birth the activity of the gamma genes is gradually replaced by activity of the beta and delta genes, the former producing about forty times as many chains as the latter. Some gamma chains continue to be synthesized, about half as many as delta chains. The manner in which these shifts take place is completely unknown.

There are several types of beta thalassemia. An individual may be heterozygous and have an abnormal allele on only one of his number 11 chromosomes in which case he may suffer a mild anemia. But if he is homozygous—has two abnormal alleles, one on each chromosome—he will be severely anemic and have very serious chronic symptoms. It is among these homozygotes that the different types of the abnormality show up most clearly. In one type, beta$^+$ thalassemia, some completely normal beta chains are produced, but many fewer than the number of alpha chains. In another, beta^0 thalassemia, no beta chains are found. In both these types delta chains are produced, sometimes in increased numbers. In a third type, delta-beta thalassemia, neither delta nor beta chains are produced. In these three types there is some increase in the number of gamma chains and hence of hemoglobin F, but this increase is never sufficient to compensate for the reduction of hemoglobin A. There is still another type in which both delta and beta chains are lacking but in which an increase in gamma chain production compensates completely for the deficit in beta and delta chains. These individuals function adequately with hemoglobin F and their abnormality is known as hereditary persistence of fetal hemoglobin (HPFH).

The diagram shows that in delta-beta thalassemia a section of DNA has been deleted extending from somewhere within the delta gene about 1,500 base pairs to the right removing most of the delta and all of the beta gene. In hereditary persistence of fetal hemoglobin there is a similar deletion, but it begins about 2,000 base pairs further to the left. The interesting deduction is that the sequences to the left of the delta gene contain instructions for turning off the gamma genes. When this information is left in, as in delta-beta thalassemia, it prevents the gamma genes from acting to replace the missing beta chains. When 2,000 more of these base pairs are removed, as in HPFH, the gamma genes remain active and compensate for the deficit. A very rare abnormality termed gamma-beta thalassemia is also known. As shown in the diagram, this allele has a deletion extending from far to the left of the G-gamma gene to a point to the right of the delta gene. A chromosome bearing this lesion produces neither type of gamma chain, nor delta chains nor beta chains and is known only in heterozygotes, for a homozygote would have only alpha chains and hence no functional hemoglobin. What is most interesting about gamma-beta thalassemia is that the complete beta gene is present on the chromosome, but in the absence of the sequences to the left of the delta gene, it never becomes active. All these observations point to the conclusion that the information for programming the activity of the gamma, beta, and delta genes lies in the nontranslated sequences associated with them. Deletions of different parts of these sequences have different effects on the programming.

But to disturb the programming it is not necessary to have a large deletion. No deletion can be detected in either beta$^+$ thalassemia or beta^0 thalassemia

(Bank, Mears, and Ramirez 1980), strongly suggesting that the programming message can be altered or destroyed by mere base pair substitutions in the untranslated portions of the DNA. This hypothesis is strengthened by a form of HPFH in which there is no detectable deletion (Tuan et al. 1980). Apparently the messages in the words of the untranslated sequences can be changed or destroyed either by rubbing them out or by garbling the letters. In those genes that have been examined there appears to be much more information in the untranslated than in the translated sequences. This suggests that there is more redundancy to be found there—more assurance for keeping the regulatory messages intact than for the translation of the polypeptides. The regulatory messages are more likely to be assailed by environmental vagaries. The channels on which they are sent and received are noisier ones than that of protein synthesis. One would expect, therefore, to find more variability from chromosome to chromosome in the untranslated sequences than in those coding for polypeptides. Evidence for this hypothesis is beginning to accumulate. A recent analysis of the human gene for insulin (Ullrich et al. 1980) found, in a sample of only four different chromosomes, two slightly different sequences. In the gene totaling 1,466 base pairs in length there were no differences in the 328 base pairs representing translated sequences, but there were four substitutions scattered among the 1,138 untranslated base pairs. Information on the sequences of the untranslated portions of genes will accumulate and it will be possible to compare such sequences in normal and abnormal alleles of the same gene. Certainly a project that cries out to be done is the sequencing of the entire gamma-delta-beta area on the human chromosome 11 for normal, beta$^+$, and beta^0 thalassemia. It will be difficult to interpret such data since we do not know how the untranslated messages are read or just how they influence gene action. But the whole question of subtle differences between alleles clearly lies in this area.

Going back now to our picture of a polygenic system controlling human pigmentation, we can visualize not merely a few dozen interacting loci, but an array of dozens of different alleles at each locus, differing subtly from each other in their contribution to the phenotype. The number of possible combinations of all such elements would be astronomical. The number of different existing genotypes would be enormous, and it is quite clear that the same phenotype could easily result from several different genotypes. Probably no two genotypes—except in the case of identical twins—would ever be exactly alike, but very similar genotypes could develop distinct phenotypes as a result of environmental influences. And pigmentation, of course, is only one character—a rather minor one, in fact. Every other functional or structural character—stature, digestion, the neural mechanism of vision—has an analogous polygenic system determining its expression and functional efficiency.

SUMMARY

What we inherit is not a set of independent unit characters such as a hooked nose, a weak stomach, or an act of armed robbery—each ready to pop up at some predetermined time. What we do inherit is a vast number of informational units coded in DNA. These units are so integrated that in the proper setting they can interact to guide the development of a new and unique individual. This proper setting includes the cytoplasm of the egg in which the process of development begins—a sort of launching pad for the informational units. The proper setting also requires an external environment consistent with the program coded in the informational units to provide further information in the form of energy and construction materials. The DNA defines the program of development that can be carried out, but alone it is inert and meaningless. Without its proper setting it is as useless as a movie film without a projector or a sound system. The information in the genetic complement of one individual gets only one complete run-through: that individual's life. If the playout equipment is inadequate, the performance will be below its ultimate potential.

4

INHERITANCE AND DEVELOPMENT OF BEHAVIOR

BEHAVIOR AS PHENOTYPE

Hyalophora cecropia is a large, showy, saturniid moth with a wing span of five or six inches. The ground color of the wings is brownish gray, but there are bold markings of white and terra cotta. In the larval state the creature is no less striking—a fleshy, naked, pale blue-green caterpillar as thick as one's thumb and from four to five inches long, baroquely decorated with rows and groups of blue, yellow, and orange tubercles bearing tiny black spines. When the larva is full grown, it stops feeding on the lilac, wild cherry, or sugar maple leaves that have been its diet, crawls restlessly around for a few hours, and then picks out a vertical extension of several inches of straight twig along which it spins a rather complicated spindle-shaped cocoon. The cocoon is made of light brown silk and is composed of an outer and an inner envelope with a loose packinglike tangle separating the two. At the upper end both envelopes have an escape hatch that will open on pressure from the inside, so that when the moth is ready to emerge it can easily escape.

If, instead of being provided with a twig on which to construct its cocoon, the caterpillar is put inside a toy rubber balloon which is then inflated, it covers the entire inner surface of the balloon with a uniform layer of silk. It produces nothing resembling the complicated structure of the normal cocoon (Van der Kloot and Williams 1953).

Female laboratory rats normally care for their young—even their first litters—in a completely competent way. They lick them, prevent them from straying from the nest, keep them warm, and give them access to the teats so that they can suckle. But female rats raised from the time of weaning with wide rubber collars similar to Elizabethan ruffs around their necks display most unmotherly behavior when, after conceiving normally and undergoing uneventful pregnancies, they produce their young. Even when the collars are removed, these rats either ignore their pups completely or treat them in utterly inappro-

61

priate ways, failing to suckle them and in many cases devouring them (Birch 1956).

Children begin to speak single words at about one year of age; by two they start joining words into two-element phrases; by four they have mastered the fundamentals of grammar and speak in sentences. Only in rare cases of extreme abnormality does a child deviate from this developmental pattern. A child born to English-speaking parents and raised in their household develops an ability to speak English. But the same child, put at the age of six months into a Chinese household in Tientsin and raised there, would never utter an English word and at four would be speaking as a four-year-old Chinese with no detectable English accent.

There is a great deal still to be learned about the complex process of inheriting behavior, but the three cases just described point up some obvious deductions that we must recognize in our efforts to investigate and understand the whole problem. There are genetic influences on behavior. Many species of caterpillars spin complicated cocoons characteristic of their species on first trial without ever having seen a cocoon or heard tell of one. Other species spin cocoons of different architecture or none at all. Female rats protect and nurse their young without a course at a prematernity clinic. Human children assimilate the rules for understanding language and for expressing themselves in speech between the ages of one and four. But caterpillars do not inherit knowledge of the architecture of a cocoon, rats do not inherit a code of maternal behavior, children do not inherit language. All three inherit the same type of unit—information sequences in the DNA which are either merely transcribed or transcribed and translated into protein to form enzymes, structural proteins, regulatory messengers, and the molecules and organelles of the protein synthesizing machinery. The products of these functional units interact with each other and with environmental factors in the process of development, and the phenotype of the individual—both its present actuality and its potentialities—results from these interactions. They take place in a matrix that includes not merely what is going on within and between the cells, but also the information in the environment as it constantly impinges on the living organism.

The caterpillar spins a cocoon because it is programmed to produce motions with its body as it spins silk which result in a cocoon when they occur in conjunction with the sensory input provided by a three-dimensional twig. In the absence of these inputs no cocoon results and the biotic program fails. In the course of normal development the female rat becomes familiar with the feel, odor, and taste of her own body from forepaws to genitalia (from licking and grooming herself). As a result of these activities she follows a program of behavior at the termination of her pregnancy which includes treatment and care of her young consistent with their survival and with the continuation of the biotic program into another generation. When deprived by the rubber ruff of the opportunity to know her own body, she arrives at the time of bearing young without ever having developed and integrated patterns of behavior that are necessary parts of the biotic program of her species. The human child inherits recipes for enzymes and messenger molecules that in the normal environment produce a nervous system capable, at a given point in development, of acting as a computer. This computer can abstract relations in the environment and relations between relations, relate these to symbols, comprehend the symbols and their interrelationships when they come as input, and produce output of similar symbols appropriately interrelated. This potentiality is peculiar to the human child and is a part of his heredity. But the child does not inherit a language. The language he speaks depends not on his heredity, but on the inputs to which he is subjected during the period when language learning is easiest. Although the two processes differ greatly in complexity, the spoken sounds around him are to the child what the twig is to the caterpillar—the necessary inputs from the environment which facilitate critical responses forming an essential part of the biotic program. In an unlimited two-dimensional universe the caterpillar achieves no cocoon; without the stimulus of spoken English words in his environment the child develops no knowledge of English; without the stimulus of any spoken language at all or in the presence of randomized nonsense the child would achieve no linguistic competence whatever.

It is extremely important to realize that behavior is as much a part of the phenotype as structure or morphology. In fact, it is rather difficult to draw a sharp line between behavior and morphology. A man has a body composed of skin, muscle, skeleton, and viscera assembled in a certain way, but in a live man posture and stance determined by muscle tonus at any given moment contribute something to the morphology which is absent from a dead man. Michelangelo strikingly dramatized this subtle but real difference in his painting of the creation of Adam, where the hand of God is vibrant with life and that of Adam is lifeless. Throughout the whole biotic cycle, morphology and behavior are constantly interacting and blending into each other. The movement of cells at gastrulation is behavior, and the resulting relative positions of mesoderm and ectoderm determine the fundamental architectural plan of the organism. Much the same can be said for the many other inductions that succeed one another as the cell masses here and there—as a result of continuing growth and displacement—trigger other bursts of cell growth and movement. Behind all this is the constant chemical activity within the cells, which is movement and behavior and has its influence on structures that become visible later. Then, too, an embryo is not a passive object. As it grows, it shows more and more evidence of its own autonomy. Heart muscle begins to pulsate before it is organized in a way that produces circulation of the blood. Other muscles twitch long before they can cause the movements that will characterize them in the complete animal. These actions are not chaotic and accidental but part of development, and a normal embryo would not result without them. The tensions of muscle tonus, for example, are necessary for the normal development of the tubercles on the bones where the muscles are attached. In both birds and mammals the embryo swallows amniotic fluid and cycles it through the body. The human fetus exercises considerably during the last weeks of gestation; and the unhatched chick constantly practices raising and lowering its head, an action that later performs the function of cracking the eggshell and allowing it to escape. In all these intricate maneuvers, events are directed by the enzymes and messenger molecules, but neither the structures nor the movements arise ineluctably from the DNA. What finally is actualized in morphology and behavior results from the interactions of these bits of information among themselves and with the environment. As events proceed, what has already happened has not only created potentialities to be actualized but has also set limits within which future development will be confined. At a given point in the development of a human embryo a group of cells is set aside to become four fingers and a thumb. This process increases the probability that the child will have a normal hand, but at the same time it reduces to the vanishing point the probability of a sixth digit. But the determination of certain cells destined to be fingers still leaves a probability that the fingers may be abnormal. Later, more detailed determinations settle the form and limits of the phalanges, the muscles, the nerves, the lines of the fingerprints and the nails. When a two-year-old child learns to say "mother," this is an achievement very likely to be retained, but a child of thirteen who knows only that word for the female parent has a low probability of ever saying *"ma mère"* in exactly the way a French child says it. Throughout the process of development the complex of future probabilities is constantly shifting.

Just as the early taxonomists thought in terms of an ideal morphological type fixed by heredity for every species, some students of animal behavior in more recent times have made very similar assumptions with respect to behavior. We hear of "species specific" behavior that is "innate." This is the old idea of a unit character produced inevitably by a gene at a given point in development. This point of view has been popularized by Lorenz (1965). Although he is careful to state (pp. 1−2), "The term 'innate' should never, on principle, be applied to organs or behavior patterns," Lorenz also says that the "concept" of the innate is "indispensable." He emphasizes a difference between innate and learned behavior. "I strongly doubt," he writes, "that the motor coordinations of phylogenetically adapted motor patterns are at all modifiable by learning" (1965, p. 71).

The stickleback is a small fish found in rivers and coastal waters in Europe and North America on the behavior of which much experimental work has been done. As the breeding season arrives, the male stickleback develops a red spot on his belly and builds a nest into which he lures or coerces a female. If another male approaches his nest, the first male drives off the intruder. A male stickleback raised in isolation from all other fish will construct a nest and attempt to drive off a dummy with a red spot on its underside if it is made to move toward the nest. This action is spontaneous on the approach of a fishlike object that is red beneath. The behavior is adaptive—is part of the biotic program of the species—and takes place even though the male has never seen another fish or red spot. The behavior must, according to Lorenz, be coded in the genetic complement and therefore be innate. To find out whether a behavior pattern is innate, Lorenz prescribes the "deprivation experiment"—raising the animal in an environment deprived of the opportunity for observing or learning the behavior. If, on presentation of the critical stimulus, the animal reacts, the pattern is innate.

Many students of comparative psychology take issue with this theory of behavior. Until his untimely death in 1968, one of the leaders of this opposing group was T. C. Schneirla of the Department of Animal Behavior at the American Museum of Natural History in New York (see Schneirla 1966). Neither Schneirla nor his group deny that the genetic complement of an animal makes possible the development of an individual capable of or predisposed to responding with a given set of neuromuscular activities to a specific stimulus. Obviously, the members of a species are likely to behave similarly. This is not the result of pure chance. Their behavior is coordinated with their morphology, which is certainly the result of natural selection. Nor can all the details of behavior have been learned from scratch by every individual. The genetic complement must contain information that produces a nervous system consistent with and adapted to the behavior characteristic of the species.

Just as morphological characters do not emerge fully formed out of the DNA but result from a complex series of interacting steps within and between cells, behavior is not read directly from the DNA. Behavior is likewise the result of complex interactions, and its potential at any stage of development is influenced by what has gone before. To achieve the potential behavior characteristic of a species, the individual must have developed in an environment within the limits tolerated by the genetic system of that particular species. And the behavior of a functioning animal is integrated; different patterns interact with each other to produce a smoothly operating whole. Consequently, there is within a species variability of behavior between individuals, even though there is a general similarity in behavior for all members of the group. To think in terms of species-specific behavior composed of "chunks" of unmodifiable patterns, as Lorenz does (1965, pp. 5, 91), is to apply the old typological concept of the species to behavior.

Schneirla did extensive experimental work on the behavior patterns of army ants and domestic cats. He concluded that the complex social system of an army ant colony and the sequence of relationships between mother cat and kittens from birth of a litter to weaning were patterned by sequential inter-actions between individuals depending on physiological functions and sensory communication. The overall pattern of interaction was genetically outlined, but the detailed events in any concrete case were determined by specific interactions between the maturation and cumulative experience of the individuals. Behavior was not "species specific" in the sense that it was a genetic straitjacket, but "species typical" in that it could vary within the limits of probability set by the genetic variation between individuals and the maturation-experience inter-actions of individual development.

In the case of a litter of kittens, birth begins a period of about three weeks during which almost all activity is initiated by the mother toward the kittens. A second, roughly equal period follows in which approach and response are mutual. During the seventh week advances from kitten to mother become dominant; the mother's failure to respond increases and weaning follows. That genetic determinants are influenced by interindividual reactions was shown experimentally in two ways. First, a kitten removed from a litter, kept in isolation in a brooder for a few days, and then returned shows the effects of deprivation of experience and is unable to adjust, as its littermates have, to the changes in the behavior of its mother during its absence. Second, the combina-tion of a female and a single kitten does not show the same three-stage development. Without the interkitten interactions, the mothers follows a much less structured pattern, resulting in a protracted mother-kitten dependence and a delayed and eccentric weaning. Thus this feline behavior appears not to be "species specific" and genetically predestined but "species typical"—"species characteristic" might be a happier term because of the association of "typical" with "typology." The behavior can be modified by differences in maturation-experience interactions during the course of development.

So, to be conclusive, a deprivation experiment à la Lorenz must demon-strate not only that the animal has not had a chance to observe or learn the reaction in question but that there is not some other experience or clue that can serve as a substitute for such observation or learning. To go back again to the female rat, experience or observation of birth or caring for pups is not necessary for normal maternal behavior. But prior experience of the smell, feel, and taste of her own body is. Lorenz (1965, p. 90) seriously undermines his case for the deprivation experiment when he says that the experimenter must exercise extreme caution to see that the animals used have not been subjected to "bad rearing." Animals that have been badly reared cannot, he says, be depended on to perform the innate species-specific behavior. It is hard to see the difference between "bad rearing" and the absence of some substitute clue of which the experimenter is unaware.

HUMAN LANGUAGE

A young fawn is able to stand and walk within a few minutes after birth. Mice are born hairless and with closed eyes and are completely helpless for several days. All mammals, however, whether they are born precocial like the fawn or altricial like the infant mouse, are not capable of surviving without a period of parental care—at least to the time of weaning. Man is not only an altricial animal but his period of infant dependency, in both absolute time and proportion of his life span, is extremely long. Man differs even more from other animals in the degree to which his behavior is guided and organized during this period of dependency through interactions with his older contemporaries. The individual gradually assimilates the way of life of the group in which he grows up. By the time he is recognized as a fully independent person, he has accepted the group's methods of providing himself with food, shelter, and clothing, of achieving satisfactory human relationships, and of reacting to the world around him in terms of goals and expectations. It is impossible to overemphasize the importance of this cultural assimilation to the individual and to the group. It is as necessary to the survival of both as the nurturing in the uterus during gestation or the providing of suitable liquid diet during infancy. Without it, the genotype could not actualize the phenotype, the individual would not become human and the species would become extinct. All of this is included when we say that the human species is characterized by having culture. This is what Aristotle meant when he said, "Man is a political animal."

Even where technology is simple, human culture is very complex. It includes practices and rules that make individual and social life possible and meaningful and which, for any given culture, constitute an integrated system. But the system did not arise as a logical plan that was consciously put into operation, and the developing child does not learn it as a set of rational propositions. Rather, he assimilates it with greater emotional than logical reinforcement. Because all human activity is developed and carried out in a cultural matrix, it is exceedingly difficult to distinguish cultural from genetic influences. In humans as contrasted with animals, the interactions responsible for the development of behavior patterns are more numerous and complex and go on for a longer time.

The central element in human culture is the human ability to form, recognize, classify, and interrelate abstract symbols and to communicate information about such activity. Usually, we think of language as spoken, but spoken language consists only of auditory symbols. Language can exist and be understood without these. The deaf learn to use visual substitutes for auditory symbols, and those who are both deaf and blind can learn to use tactile ones. The essence of language is the ability to place cognitive units in the relationship of subject, predicate, and modifiers; the choice of symbols used to transmit information about such relationships is of secondary importance. In spite of early failure to elicit speech from a chimpanzee, numerous careful and painstaking experiments have been carried out in the last two decades designed to teach chimpanzees linguistic communication by means of American Sign Language, a system of plastic tokens, or the keyboard of a dispensing machine (Terrace et al. 1979, Savage-Rumbaugh, Rumbaugh, and Boysen 1980). The result has not been a clear-cut answer to the question "Can apes use language?" There is disagreement among the investigators. Apes can learn to associate symbols with objects and forms of behavior and can build up impressive vocabularies. But evidence that they can arrange and organize these symbols in accordance with rules of syntax is very meager. The gulf between the painfully slow progress of the chimpanzee, even under the best expert instruction, and the virtuosity demonstrated by the prattling three-year-old child is sufficiently great to brand the chimpanzee as linguistically incompetent.

Linguistic competence is a trait so fundamentally human that no population exists without it, and within populations the number of linguistically incompetent is extremely small, and these incompetents display other obvious abnormalities. Within the competent group there are differences in achievement varying from limited facility of expression to conversational virtuosity, but even individuals with severe speech or hearing defects attain functional communication. Whatever the genetic component of linguistic competence, it functions in human populations with a very low frequency of failures.

Between human populations there are, of course, vast differences of speech. The number of mutually incomprehensible languages is in the thousands. There is probably no character for which there is a higher correlation between parent and offspring than language. But on so many occasions it has been possible to observe the effects of raising a child born to parents speaking one language, in a household where another was spoken, that few people have any doubt that any child will learn with equal facility whatever language he hears spoken around him. Yet literature is full of fantastic notions to the contrary, from the pharaoh Psammeticos of the Twenty-sixth Egyptian Dynasty who, according to Thucydides, had two children raised in isolation on the assumption that they would speak to each other in a mythical original language, to C. D. Darlington in twentieth-century England, who argued (1947) that carriers of the gene for the O blood group found it easier than other people to make the lisping sound of English "th." But it seems incontrovertible that while man's genetic system makes him capable of developing linguistic competence, it in no way determines what language he will speak.

There is a growing consensus among students of linguistics that any living language has the potential of enabling those who speak it to express whatever ideas they are capable of formulating. All languages are constantly changing in vocabulary and grammar; and every language is strongly influenced by the activities and interests of the people who speak it. Classical Greek had no vocabulary for discussing French literature; Eskimo languages are poor in terms relating to tropical plants; and few educated people of today can discuss the intricacies of medieval heraldry. A language spoken by a small, isolated population with primitive technology will be provincial in that its vocabulary will be limited to local objects and to concepts of local interest. A large population with a complex division of labor, an intricate technology, and many contacts with other populations develops a more versatile language as a result of the many stimulating challenges to which it is constantly exposed. But any functioning language has all the potentialities of human linguistic competence. The provincial language of a small, isolated, technically primitive group is not primitive in the sense that it is nearer to some prehuman or subhuman system of communication. It is a manifestation of the human linguistic potential, the ability to analyze the environment in terms of relations, code this analysis by the use of symbols, and then use the symbols to make statements about relations between relations. Whatever changes have occurred in this linguistic potential during the course of human evolution appear to have occurred for the whole of the human species. If the Bushmen of the Kalahari Desert were to increase to a population of 100,000,000 and advance technically to a position equivalent to that of western Europe, there is no reason to believe that their language would not adjust to the new situation. A great many new terms would have to be invented and a great many loan words would probably be incorporated, but as a vehicle for expressing relationships their language is potentially the equal of Chinese, Russian, or English.

Another characteristic of language of great interest to the student of human behavior is the difference in the facility with which it is learned before and after puberty. By the age of four or five a child has usually mastered the fundamentals of the language to which he has been exposed. Children hearing two languages usually learn them both without confusing them. Up to the age of ten or eleven a child can assimilate a second or a third language with much the same ease that he learned his first, speaking without an accent and without grammatical clumsiness. From the age of twelve on, learning another language not only becomes progressively harder but the very process of learning changes. The unconscious assimilation characteristic of the young child is replaced by a conscious effort. Pronunciation becomes a task, vocabulary rote learning, and grammar the mastering of a new set of rules. The immigrant who comes to a new country before the age of eleven learns to speak the new language like a native. As a young adult he will not be recognizable as a foreigner either by accent or by locution. But the immigrant of eighteen or nineteen will probably never speak without a detectable accent. One who arrives in his mid-twenties is likely to retain a strong accent and never be completely at home in his new language, especially with regard to the use of prepositions, the subtleties of word order, and the use or omission of articles.

Although the period of great facility for acquiring languages extends over many years before it fades away, in certain respects it strongly resembles the process of imprinting. The common graylag goose of Europe, when it hatches from the egg, will follow any somewhat taller moving object that emits rhythmic sounds. Normally, of course, a somewhat taller moving object emitting rhythmic sounds is the mother goose and it is clearly adaptive for the gosling to follow her. Within a few days the gosling has achieved a fixation on its mother and will follow no other moving object—will accept no substitute. But, as Lorenz (1965, p. 56) has demonstrated and so amusingly described, a newly hatched gosling can commit itself to following a brooder equipped with a buzzer or a person who stoops low and imitates a goose, and the commitment to the brooder or the person becomes just as permanent as that to the goose even though it is ridiculously nonadaptive. Imprinting in the gosling is very dramatic because it takes place very rapidly and does not seem to involve conditioning or learning. The gosling comes to follow the goose efficiently and faithfully without any reward for performance or punishment for wandering. This sudden development of a behavior pattern is characteristic of precocial animals where it is of great adaptive value for the young to adopt a stereotyped relationship to the parent quickly, very soon after birth. The kitten and the mother cat arrive at an equally adaptive relationship, but in the seclusion of the nest this can develop more slowly.

What is similar between imprinting and human language development is that in both the activity develops spontaneously and with a minimum of external pressure during a special period. If the behavior is not achieved during this time, it can be attained later only as a result of high motivation whether the sanctions are internal or external. During the critical period there is a competence in the developing organism which can actualize as a behavior pattern, given the proper experiences. The details of the actualization will be enormously influenced by the experiences. As the end of the critical period approaches, the probability of actualizing the potentiality becomes less and less, no matter what the experiences.

It was not so many years ago that race and language groups were confused and even identified by respectable students of human affairs. In 1899 W. Z.

Ripley published *The Races of Europe*, in which linguistic characters were the primary determiners of racial groups. His "races" were Latin, Slavic, Teutonic, and so on. Today it is realized not only that language is not inherited but that whole populations sometimes change language because of political or other cultural pressures. The population of ancient Gaul gave up its Celtic languages and adopted Latin; the pre-Greek languages of Anatolia were replaced by Greek in the Hellenistic period, and then in the Middle Ages Greek gave way to Turkish. That two people have the same native language says nothing about common descent. A Georgia black, a Hawaiian of Japanese ancestry, and an inhabitant of London all speak English, but their biological origins could scarcely be more diverse. It is quite apparent that language and race are completely independent and that if racial groups differ genetically in linguistic potential, the difference must be very slight—so slight as to be inconsequential in practical communication.

INTELLIGENCE

There are still people who believe that individual personality traits and whether one's behavior is vicious or virtuous are determined in a simplistic manner by genes directly specifying such traits. But today there is much less talk of genes for alcoholism or criminality than there was even thirty years ago. To some extent this is probably the result of the difficulty in imagining a DNA sequence that translates into embezzlement or homicide. The behavioral character still commonly assumed to be rigidly determined by a one-to-one relationship to one's alleles is intelligence. Arguments to prove genetic differences in intelligence between races are constantly being put forward in bars, in faculty dining rooms, in newspapers, and in scientific journals. This is probably the most controversial subject in the entire field of race relations.

It is important to distinguish the operational definitions of intelligence provided by the educational psychologists from the loose thinking about the concept which is widespread and by no means confined to people of limited education. In conversation over cocktails or at dinner parties, "intelligent" is commonly used to mean "like me" or perhaps even more accurately "as I like to picture myself and wish I really were." While there is much inconsistency and ambivalence in popular thinking on the subject, to many laymen intelligence is that quality that results in one's doing the proper thing, being good company, and being easy to get on with. Lack of intelligence causes people to be troublesome, bizarre, unhygienic, and to espouse absurd causes. A generation or two ago when immigrants from Europe and their first-generation offspring formed a large proportion of the population of the United States, the slightly more assimilated used to express their scorn for the later arrivals who had not yet

71

learned to hide their old country origin by calling them "ignorant." The word was spoken with explosive disdain. During World War I, Loredo Taft, the Illinois sculptor, began a lecture on Medieval French art to an audience of American doughboys in Paris with this tongue-in-cheek sentence, "There was once a good and popular king of France whose subjects called him *Jean le Bon* because they were exceedingly ignorant and could not speak English." "Intelligent" as commonly used today is the antonym of "ignorant" in the former usage.

In fact, being intelligent has become the primary virtue, corresponding to being wellborn in the eighteenth century and God-fearing in the nineteenth. Poverty, crime, disease—unpleasantness in general—flow from lack of intelligence. Affluence, success, attractiveness, and propriety are bestowed on the intelligent. It is mistily postulated by the man in the reclining chair that a world inhabited only by Einsteins, Beethovens, Schweitzers, and Newtons (Isaac, not Huey) would have no serious social problems. Those who extol intelligence in this fashion usually pick dead or remote intellectual heroes, whom they would undoubtedly have found very disappointing to deal with directly in everyday life.

If we go to the educational psychologists, we get quite a different concept of intelligence. Most would agree, I think, that intelligence is an ability to recognize abstract symbols and their meaning and to classify them and reclassify them in different relationships. This ability can be measured by certain tests. Since the ability increases from infancy to the late teens, the intelligence quotient (I.Q.) was originally defined as the ratio ($\times$ 100) of the individual's score to the average score for his age group. Thus an individual having the average score for his age group had a ratio of 1.0, or an I.Q. of 100.

Measurement of the I.Q. is now carried out by administering standard tests—some single, some in a battery with varying content: verbal, mathematical, pictorial, and diagrammatic. These tests have been calibrated by trying them on carefully selected representative groups of subjects and adjusting them until they give a normal distribution of scores. The scores are then standardized and the results tabulated, so that from the raw score of a subject and his age one can read his I.Q. from the table. Since 100 is the population mean, it is the center of the distribution. The standard deviation is set arbitrarily at 15, which means that 66 percent of the whole distribution lies between 85 and 115, while 50 percent of the distribution lies between 90 and 110.

Intelligence tests have been developed and used most extensively in the United States and Great Britain, primarily as a means of predicting the achievement of children in school. Since academic achievement is a goal rated very highly by members of the middle class, persons of middle-class origin as a group usually score higher on the tests than members of other social classes such as blue-collar workers, migrant laborers, or depressed minority groups. A cursory reading of representative tests is enough to reveal that they are redolent of middle-class atmosphere and values. The Wechsler Adult Intelligence Scale (WAIS) (Wechsler 1955), one of the tests now widely used, asks in its section on comprehension: "Why are child labor laws needed?" According to the manual of instructions for persons administering the test, one must, to obtain full credit on this question, mention at least two of the following reasons: "health, education, general welfare, exploitation, avoid cheap labor." If one answers by saying that child labor laws serve to keep children from competing with adults for jobs, one is given a score of zero. The whole question is considered from the point of view of an economically secure person who can afford to view children as precious social assets to be cherished and protected. Anyone whose daily struggles may cause him to evaluate children differently is likely to score lower on the WAIS.

Jensen (1980, p. 4) argues that one cannot evaluate a test by questioning the "face validity" of specific items. Clearly, the test asks for more than the ability to recognize certain facts. But anyone perusing either the WAIS or the Stanford-Binet Intelligence scale cannot help concluding that the fair-haired testee whom the selector of the items must have had in mind while constructing the test bore a closer resemblance to Tom Swift than to Huckleberry Finn. The constant vulnerability of the test maker in an unstable world is illustrated by the item in the 1960 Stanford-Binet test which asks for "two reasons why there should be plenty of railroads in the United States."

There is considerable disagreement among psychologists on just what the scores from the I.Q. tests and their distribution mean. One group, following C. Spearman and Sir Cyril Burt, is convinced that the I.Q. is a measure of a unitary ability that exists at a given level in an individual and cannot be resolved into components. It is often referred to as g (general intelligence). Other psychologists, following L. L. Thurstone, see intelligence as composed of several special abilities that can exist in an individual in different proportions. This point of view has been carried furthest by J. P. Guilford (1967), who analyzes intelligence in a three-dimensional continuum with five operations along one axis, six products along another, and four contents along a third. This gives 120 boxes, or cells, each representing a separate component. He maintains that tests can be constructed which measure performance on a single component and are substantially independent of the others. The battery of tests used to measure the I.Q. appears to recognize different facets, if not components, of intelligence. It can be divided into the verbal tests and the nonverbal, and there are systematic differences between the scores in the two types. For example, girls consistently score higher on the verbal tests, boys on the nonverbal. Psychologists who see intelligence as unitary base their argument on the degree of correlation between the scores of different tests and they contend that this correlation reflects the unity of g. Psychologists who favor the theory of independent components maintain that the correlations between tests are the result of questions or tasks that overlap different components, and that careful construction of tests can reduce intertest correlation to the vanishing point. This whole problem is extremely complex and involves formidable statistical and psychometric techniques. The plain fact is that there is substantial disagreement among the experts. As L. Willerman (1979, p. 103), puts it, "it is apparent that intelligence is not yet well understood." This greatly complicates any attempt to come to an understanding of the role played by intelligence in human populations.

A second disagreement between psychologists concerns the process of the development of intelligence in the growing individual. This is a recrudescence of the old question of the relation of genotype to phenotype. Intelligence as measured is a phenotypic character. To what extent and in exactly what way is this character determined by the genotype? The point of view that emphasizes genetic determinism is ably expounded by Professor Jensen (1969a). He recognizes that phenotype results from an interaction between genotype and the environment, but he pictures the genotype as determining an upper limit of phenotypic expression which cannot be exceeded. Environment can depress the expression to a point below this upper limit, but unless the environment is severely substandard, the upper limit will be reached. Hence most of the difference in I.Q. scores between any two individuals is genetically determined.

Most developmental geneticists would question the rigidity of this assumption. They would picture instead an upper limit that would include a certain percentage of such genotypes, say 95 percent, but the other 5 percent would be found above the limit. A somewhat higher limit would include 99 percent, but 1 percent would exceed even this. They picture no absolute upper limit, only a series of limits defined by different probabilities.

A point of view regarding the development of intelligence in the individual in harmony with these concepts of developmental genetics has been outlined and defended by J. M. Hunt in *Intelligence and Experience* (1961). This position is essentially Schneirla's theory of sequential interactions in cumulative experience applied to the development of human intelligence. Hunt and his school emphasize the continuous interaction between the growing child and his experience, pointing out that today's experiences impinge on the child of today and that what has happened to him previously is as much a part of him as his genotype. One can accept the idea that the gene pool of the species predisposes the members to species-typical behavior and that individual genotypes set probabilistic limits to individual behavior and at the same time maintain that the behavior of the individual is substantially influenced by interactions between organism and environment during development.

In mammals generally, and especially in man, there is substantial organic maturation of the brain after birth. The precise extent to which this cerebral development can be adversely affected by the environment is not known, but there are reasons for believing that its normal course is dependent on a fine balance in nutrition. In experimental animals, protein-deficiency in early development produces abnormally small and underdeveloped brains; and animals that have suffered such deficiency are less competent at problem solving than those that have been well nourished (Eichenwald and Fry 1969, p. 646).

In man there is a recessive allele that when homozygous produces the condition known as galactosemia. Galactosemic babies, who fortunately are very rare, are very sickly. If they survive infancy, they develop severe symptoms including cataracts, a grossly enlarged liver, and severe mental retardation. The abnormality in this disease is an inability to metabolize galactose, one of the sugars found in milk. If a galactosemic infant is diagnosed within a few days after birth and given a diet free from galactose, it will develop normally—no cataracts, no liver damage, and no metal retardation.

The inability to metabolize galactose causes a derivative of the sugar—galactose-1-phosphate—to accumulate in the blood. Although small amounts of this substance are found in the blood of all normal persons after they have ingested galactose, when it reaches the abnormal levels found in galactosemics, it produces toxic effects and causes, among other symptoms, severe mental retardation. There are other hereditary metabolic diseases in which an abnormal level of a normal metabolite produces a mental defect and in which the defect can also be prevented from developing by proper therapeutic diet. One must conclude that a potentially normal nervous system may be caused to develop defectively by abnormal levels of a usual metabolite. It seems reasonable to suspect that certain combinations of slightly abnormal alleles may result in less drastic mental impairment than that produced by untreated galactosemia and may account for some of the scores below the mean in the I.Q. distribution. But it seems just as reasonable to believe that external factors such as unbalanced or deficient diet or viral infection may react with the genotype to produce a low I.Q. as a phenocopy.

To date we have no precise information on the effect of nutrition on the development of intelligence in those genotypes potentially within the normal range. But our lack of knowledge is no reason for assuming that the influence of nutrition is negligible; there are many reasons for thinking that it must be substantial. We lack knowledge on this question because it is extremely difficult to set up the kind of study that would give us the answer. Ideally, such an experiment should have a large number of cases where nutrition was varied according to a precise plan and at the same time all other environmental influences and the genotypes were just as carefully randomized. With humans, this is impossible to do. A careful study by Winick, Meyer, and Harris (1975) compared height, weight, I.Q., and school achievement among three groups of Korean orphans adopted by families in the United States. On the basis of medical records of their infancy in Korea, the children were divided into three groups: malnourished, moderately nourished, and well nourished. After several years with their adoptive families, the three groups scored in the same rank order as their original nutritional status. For height and weight all three groups were above the 50th percentile for Koreans but below that for Americans. For I.Q. and scholastic achievement all three were above the 50th percentile for Americans. How is one to disentangle nutritional from genetic and from other environmental influences? And what, precisely, does nutrition mean in this case? Studies designed to investigate these questions are constantly being undertaken, but because of the inherent difficulties, it will be no surprise if an unequivocal answer is slow in coming.

It is also argued by many psychologists that the type of experience that a developing animal undergoes in its early life has a lasting effect on the degree of competence in behavior that it displays as an adult. There is a very considerable body of data from experiments with animals to support this contention. Laboratory rats, for example, which have been raised in a varied environment score consistently higher on problem-solving tests than littermates reared in cages where environmental stimulation was minimal in amount and variety. There is also evidence that suppression of sensory experience actually affects the physical maturation of neural tissue. Chimpanzees raised in total darkness for a year and a half do not develop normal vision when exposed to light, and their retinas and optic nerves have been found to be histologically abnormal. There is even evidence from kittens and mice that deprivation of light stimuli to the eye produces detectable changes in the nerve cells of the portions of the brain associated with visual perception (Hunt 1969, pp. 137–140, Muir and Mitchell 1973).

The occurrence of highly important developmental processes in the human central nervous system in the early years after birth and the drastic decline with full maturation of certain human learning abilities, linguistic, for example, together with data from animal experiments, strongly suggest that full actualization of the behavioral potentialities of a given genotype may be greatly influenced by experience during development. To accept this possibility is not to deny the influence of the genotype. It is merely a recognition that the behavioral phenotype, like the morphological, is the result of the interaction of genotype and environment. Here again, as in the influence of nutrition, the precise effect of experience on the development of intelligence is not known because of the difficulties encountered in designing studies to measure it. There is no evidence that it is negligible, and there are good grounds for believing that it is substantial. This point of view underlies the rationale of programs of compensatory education. The chief difficulty in trying to translate it into action is that we know so little of the details of the process through which either nutrition or experience may influence the development of behavior that we can only guess at when and with what we should attempt to compensate.

Since there is agreement on neither the nature of intelligence nor the influence of experience during its development, it is not easy to deal with the question of the heritability of the I.Q. Heritability is a concept that has been carefully worked out by mathematical geneticists and which plant and animal breeders have found very practical for predicting the results of selection programs. Heritability can have a value between zero (0.0) and one (1.0), and this figure tells the breeder what proportion of the difference between the mean phenotype of his selected parents and the mean of the population from which they were selected can be expected to appear as an addition to the mean of the offspring of the selected parents. Mathematically, heritability is defined as the proportion of the total phenotypic variance in a population attributable to genetic influences.

There are several ways of obtaining estimates of heritability. With domestic plants or animals these all involve controlled crosses in which genotype, phenotype, and conditions of rearing are followed as carefully as possible. The resulting data are subjected to statistical analysis involving a varying number of assumptions depending on how much information previous analyses of the lines and the characters have provided. But however carefully estimated, heritability is a characteristic not of an individual nor of a trait but of a particular population in a given environmental situation. Heritability makes a prediction about the relationship between the population mean for a character, the mean of a set of parents drawn from that population, and the mean of the offspring of those parents. For a character to show heritability, different alleles affecting the character must be segregating. If a population is completely homozygous for the alleles influencing a given character, that character will have in that population a heritability of zero. In such a population, however, the variability will not necessarily be zero, but it will be nongenetic, that is, environmentally produced. Heritability measures the amount of *genetic difference* available. Hence the measurement of the heritability of a trait in one population gives no information on the heritability of that trait in some other population.

Theoretically, it is possible to partition total phenotypic variance of a population into a genetic component (heritability, which can be further subdivided), an environmental component, and a component arising from the interaction between the genetic and the environmental. For data on human intelligence the environmental factors, although unquestionably of importance, are so poorly understood as to detail and so refractory to quantitative measurement that it is questionable whether going beyond a simple partition into a genetic and an environmental component is worthwhile. In attempting to apportion the variance for differences in intelligence into these two parts, it is not unreasonable to contend that since 1.0 is the highest possible value for heritability, the value of one minus the heritability must represent that proportion of the total phenotypic variance attributable to nongenetic influences. If, in a given population, heritability for a trait is found to be 0.7, the 1.0 - 0.7 or -0.3 of the total phenotypic variance must result from nongenetic factors. This does not mean, however, that environment can contribute only 30% of the variance. Under some other environment the same population might show a lower heritability and a higher nonheritable component. The figure 0.3 in this illustration means only that if all nongenetic influences on the phenotype had been eliminated, the total variance would have been reduced by 30%. In a population with much genetic diversity and little environmental variation, heritability will be high. In a population with little genetic diversity, heritability will be low and environmental influences relatively greater. If a successful breeding program changes the mean of a character and increases the genetic uniformity of the population, it also reduces the heritability of that character.

Since we cannot make controlled crosses in human populations, for making estimates of heritability we are forced to use phenotypic data collected from matings that have occurred in some given population. Estimates of heritability of quantitative traits can be made from correlations for such traits between relatives. Such correlations should decrease as one goes from a close to a more distant relationship. Table 4 lists a set of correlations of this kind. A method has been worked out for measuring objectively the number of fingerprint ridges on each finger. The table contains correlations between different classes of relatives of the total number of ridges on each individual's ten fingers (Holt 1961). Column 1 gives the types of relatives compared, column 2 the proportion of

TABLE 4. Correlations of a Quantitative Trait between Classes of Relatives

Relationships	Genes in common	Finger ridge count correlations	Heritability
Identical twins	1.00	0.95 ± 0.01	0.95
Fraternal twins	0.50	0.49 ± 0.08	0.98
Sibs	0.50	0.50 ± 0.04	1.00
Mother-child	0.50	0.48 ± 0.04	0.96
Father-child	0.50	0.49 ± 0.04	0.98
Mother-father	0.00	0.05 ± 0.07	0.00

SOURCE: Holt (1961).

genes they have in common, column 3 the observed correlations, and column 4 the heritability—the correlation divided by the proportion of genes they have in common.

For all combinations except mother-father the heritability is substantially 1.00. For parents, who presumably have no genes of common origin, it does not differ from zero. Human finger ridge counts appear to be determined almost completely by a polygenic system with little environmental influence. It is interesting to note that, in contrast to this extremely high heritability, figures for the characters for which selection is commonly carried out in domestic animals rarely approach these levels. In cattle, heritability for milk production is about 0.30, for staple length of wool in sheep, 0.25, and in swine, for body weight at corresponding ages, about 0.30. In poultry, out of twenty-six characters listed by Lerner, only three approach 0.80 heritability. These are all types of egg quality: weight, shell color, and albumin content. Of six characters relating to egg production, none reaches 0.50 (Lerner and Libby 1976, pp. 184—85).

Figure 7 shows a plot of a large collection of correlations of I.Q. scores between various groups of relatives gleaned by Erlenmeyer-Kimling and Jarvik (1963) from fifty-two different studies. These range from 0.00 for unrelated persons raised apart to over 0.90 for monozygotic twins raised together. For most categories the observations cover a considerable range; the median for each is shown by a short vertical line. Since these are correlation values, they must be divided by the proportion of genes in common to give the heritability. For parent-child, siblings, and dizygotic twins the median correlations are close to 0.50. Since in all three of these categories the individuals compared share half of their genes, the heritability in all three is not significantly different from 1.00—a startling figure, more like that for finger ridge counts than those for milk production or length of wool fiber.

Are we to take such heritability figures at face value? There are several reasons for caution and skepticism. First of all, there is the question of the measurements themselves on which the correlations are based. Numbers of finger ridges, color of egg shells, thickness of back fat on swine, concentration of sucrose in beets can all be defined physically or chemically and measured objectively. The I.Q. is not the result of a physical or chemical determination. It arises as the result of behavior—of satisfactory or unsatisfactory answers or performances on a set of questions or tasks.

These questions and tasks are neither directly designed nor finally selected with a clear and precise understanding of exactly what they are going to measure. In constructing a test, potential items are first carefully scrutinized by an "item writer" or a panel of several judges. Next, the acceptable items are tried out by giving the whole collection as a test to a large number of subjects

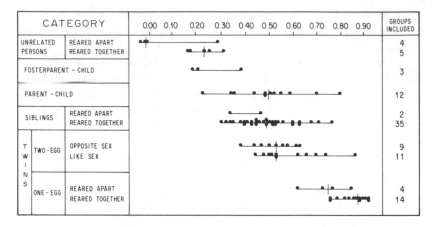

FIGURE 7. Correlation coefficients for "intelligence" test scores from 52 studies. Some studies reported data for more than one relationship category; some included more than one sample per category, giving a total of 99 groups. More than two-thirds of the correlation coefficients were derived from I.Q.s, the remainder from special tests (for example, Primary Mental Abilities). Midparent-child correlation was used when available, otherwise mother-child correlation. Correlation coefficients obtained in each study are indicated by dark circles; medians are shown by vertical lines intersecting the horizontal lines that represent the ranges. (Reproduced from Erlenmeyer-Kimling, and Jarvik 1963, *Science* 142:1477.)

typical of the population for which the final test is intended. From the results of such tryouts, the items can be evaluated as to level of difficulty, correlation with the test as a whole and numerous other characteristics. On the basis of this information a final test can be constructed which is then given to a large, random sample of a specified population and standardized for that population, that is, given a system of scoring that will result in a normal distribution of scores, the mean of which is given a value of 100 and the standard deviation set at 15 (Jensen 1980, pp. 74–75, 136–38).

These details are important because they indicate that the scale for measuring the I.Q. is the result of a complex interaction between the test makers and the population to be tested. It is assumed that the distribution is normal. Items are finally picked and the scoring adjusted to make this so. There may be subpopulations that differ. Some items show such a difference between the sexes and some tests are deliberately adjusted to compensate. Subpopulations that are not recognized are lost in the homogenization of normalization and standardization. Because of these adjustments an individual's score can have meaning only on the assumption that he is in fact a member of the population on which the test was standardized.

Because of this involuted circularity in the construction of the I.Q. scale, how can we know that it is independent of the influences that may affect the relation between genotype and phenotype? It is essentially this question that has been raised by two recent papers questioning the whole corpus of studies on the heritability of the I.Q.: one by an astronomer, David Layzer (1974) and another by two biologists, M. W. Feldman and R. C. Lewontin (1975).

Two of the influences that spring to mind as candidates for influencing the I.Q. are culture and environment. Much has been written by the psychometricians on the culture-fairness of tests. In general, they seem to equate culture with language so that nonverbal tests are assumed to be less culture-bound than those based on words. Jensen, in *Bias in Mental Testing*, argues that language differences cannot be the reason for the lower mean of black I.Q. scores because blacks do better on the verbal tests than on the nonverbal. Of course, one cannot do one's best on a test administered in a language in which one does not feel at home, but culture is more than language alone. Cultural anthropologists have long been aware of the subtle influence of culture on the habits of thought and feeling and on the ways of reacting to stimuli. This point was ably presented almost a half century ago by Ruth Benedict (1934) in *Patterns of Culture*.

Since then much work has been undertaken by both anthropologists and psychologists on cross-cultural differences between both widely divergent cultures in different parts of the world and different socioeconomic levels within similar Western cultures in the United States and Great Britain. Bernstein (1971), for example, found that for young children of lower socioeconomic status language is learned as a much more restricted method of communication as compared with the role it plays for children of upper socioeconomic level, and that toys among the former are pacifiers to keep children from bothering busy mothers, whereas among the latter toys are instruments for investigation and learning. Deutsch (1965) calls attention to the absence of encouragement or recognition for the child of lower socioeconomic status who shows cognitive or linguistic achievement.

Cultural influences are subtle, hard to define, and harder still to quantify. The same spoken or written word may have two sets of meanings and emotional overtones for two individuals of different cultural backgrounds. The two sets will probably overlap and partially coincide, but the nonoverlapping portions will tend to generate differently modified responses. Similarly, reactions to abstract forms and patterns cannot be completely free from culturally influenced emotional flavor. When viewed with these considerations in mind, the notion of a culture-free test becomes extremely hard to accept. In spite of all the work that has been done, there has been no effective integration of psychometrics and cross-cultural studies.

As Vernon (1979, p. 124) observes, "white middle- and upper-class children, on first arrival at school, differ fundamentally from lower-working-class children and especially from those children who are further differentiated by ethnic or racial origins, such as blacks, Chicanos, and American Indians in the United States, or West Indians, Indians, Pakistanis, and Cypriots in Britain. Middle-class children are advantaged not merely in such surface characteristics as better clothing or a different speech accent; they are also much more fluent and grammatical in expressing ideas, they have had a lot of experience at home in school-type tasks, and they are generally more cooperative with teachers and accepting of school aims; they will, therefore, settle down to learning more readily."

It is hard to draw a clean line between culture and environment, for culture is a continuing, established, environmental system. Nevertheless, within a given culture or socioeconomic level different individuals have different experiences, different associates, and different patterns of relationship with these associates, and gradually develop individual patterns of behavior that we recognize as part of the personality. It seems reasonable to hypothesize that this group of influences is not determined entirely by the genotype and that it may affect the phenotypic expression of the I.Q. In estimating heritability of characters in domestic animals the breeders make every effort to randomize environmental influences. Lerner (1972, p. 412) has pointed out that estimates of heritability may be doubled or halved depending on whether environmental factors are randomized or left uncontrolled.

Studies of heritability of human intelligence are probably most deficient in failing to randomize environmental factors. But this is a formidable task, chiefly for the reason that no one knows what the factors are. Socioeconomic status has been most often invoked as the most important environmental variable, but it turns out to be remarkably difficult to define it precisely—whether by income, employment status, area of residence, or years of schooling—and, however defined, it does not represent environmental uniformity (Vernon 1979, p. 117). Father's income or bank balance, the number of books in the house, the census tract in which the family lives, what college mother attended—none of these directly affects the child's I.Q. What does influence it must be the cumulative effects of subtle experiences through time. That the I.Q. is influenced by birth order and by the mental age of the other family members present when a child is growing up (Zajonc 1976), strongly suggests that there are important environmental factors still to be identified.

Although the Erlenmeyer-Kimling and Jarvik compilation suggests a higher heritability of the I.Q. from studies of sib-sib and parent-child correlations than from those based on twins, twin studies have been more often cited as evidence for a high heritability of the I.Q. The psychometricians in general have been hereditarian, convinced that intelligence was innate. Francis Galton, their founder, entitled his book *Hereditary Genius*. C. E. Spearman and Sir Cyril Burt in England, L. M. Terman, J. M. Cattell, R. B. Cattell, and A. R. Jensen in the United States have been of the same persuasion. L. S. Hearnshaw, Burt's biographer, writes that the innateness of intelligence "was for him almost an article of faith, which he was prepared to defend against all opposition, rather than a tentative hypothesis to be refuted, if possible, by empirical tests" (1979, p. 49). Burt's studies of identical twins raised apart supported the hereditarian position. In fact the figures from his 1966 paper formed the strongest single argument for the widely quoted value of 0.8 as the heritability of the I.Q. Jensen used Burt's heritability figures based on fifty-three pairs of monozygotic twins raised apart in a 1970 paper in *Behavior Genetics* (1:133—48) in which he combined them with figures for sixty-nine other pairs taken from three other studies and came up with a heritability figure of 0.824.

In October 1971 Burt died. In April 1972 Dr. Leon Kamin of the Department of Psychology of Princeton University raised questions as to the soundness of Burt's empirical data at a departmental colloquium there and he continued his criticisms at various academic meetings during the following year. These he elaborated in *The Science and Politics of I.Q.*, published in October 1974. Kamin cited vagueness and inconsistency in statements concerning the methods of testing and of collecting data and incredible numerical identities in correlations based on different sample sizes as the number of observations increased from the 1955 paper through the succeeding papers of 1958 and 1966. In March 1974 Jensen published in *Behavior Genetics* (4:24) an evaluation of the figures that Kamin had questioned and concluded that in view of the vagueness of the sample sizes "the correlations are useless for hypothesis testing." In October 1976 the *Sunday Times* of London published an article by Oliver Gillie accusing Burt of deliberate fraud in fabricating data to support his theories. Since then a controversy over Burt's integrity has raged, not only over the twin data but over other studies as well. There are those who argue that he was guilty of no more than carelessness. Hearnshaw began work on the biography in 1972 with no thought of the possibility of fraud. In the completed work he is unable to acquit Burt of deception. There is no question that these revelations have seriously tarnished the reputation of the hereditarian position and given support to those opposing the use of I.Q. tests for categorizing students in the schools as unfair to minorities.

So strong has been the reaction against I.Q. tests that during the 1970s several of the school systems in larger cities in the United States—including Chicago, Houston, Los Angeles, and New York—have discontinued them as a means of classifying and placing children (Jensen 1980, p. 34). In October 1979 the United States District Court in northern California ruled that the use of I.Q. tests for classifying blacks as "educationally mentally retarded" was a violation of their constitutional rights (*New York Times*, Oct. 17, 1979: D, 23). On August 8, 1980 the Secretary of the Army ordered personnel officers to remove from field records the scores of the Armed Forces Qualification Test. This followed a recommendation by a departmental study group that the Army do away with this test, which had come to be regarded as an intelligence test, and replace it with a procedure designed to measure ability to do a job rather than position with respect to an abstract norm (*New York Times*, Aug. 9, 1980, p. 5).

Whether Burt did or did not falsify his data, does not, of course, change the heritability of the I.Q. The hereditarians cite other studies supporting their position, although no other single study gave such clear and persuasive backing. The question is: Why was Burt not challenged earlier? All the discrepancies that Kamin began pointing out in 1972 had been available for inspection in published papers for at least six years. One of these papers was Burt's Walter Van Dyke Bingham Memorial Lecture delivered in London, May 21, 1957, and published the following year in *The American Psychologist* (13:1—15). The published version contained the following: "The figures show that the abler children of the working classes, even when they have obtained free places or scholarships at secondary schools of the 'grammar' type, frequently fail to stay the course: by the time they are sixteen the attractions of high wages and of cheap entertainment during leisure hours prove stronger than their desire for further knowledge and skill, and easily overcome their original resolve to face a long prospect of hard sedentary work in *statu pupillari*. . . .

"Underlying all these differences in outlook, I myself am tempted to suspect an innate and transmissible difference in temperamental stability and in character, or in the neurophysiological basis on which such temperamental and moral differences tend to be built up. Tradition may explain much: it cannot account for all."

This startling statement is hard to reconcile with the objective attitude assumed to characterize a scientific investigator. It is fair to paraphrase it: "Even when a member of the lower classes has the genes for a high I.Q., they do him no good because he also has genes that make him defective morally and prevent his becoming a gentleman."

This public expression of prejudice did no harm to Burt's standing with the hereditarians. Jensen was present at this lecture and in January 1972 he wrote in the preface to *Genetics and Education*: "His lecture was impressive indeed; it was probably the best lecture I ever heard and I recommend it to all students of psychology and education. (It was published in *The American*

Psychologist 1958, 13:1—15)." Herrnstein, in a letter of July 16, 1973, dealing with the Burt controversy and quoted by Kamin in *Science* (1977, 195:247), wrote: "Until Kamin started his campaign to discredit Burt, there was no hint that Burt was suspect."

Alexandre Dumas, *père*, once made the observation: "There is nothing more convincing than a deep conviction." It is hard to escape the conclusion that it never consciously occurred to the hereditarians to look for vagueness or inconsistencies in Burt's data or to be disturbed by openly expressed class prejudice because the studies supported a deep conviction long since accepted. It is not what Burt did but how widely and uncritically his work was endorsed that has increased the amount of skepticism concerning the heritability of the I.Q. The uneasiness that this development has caused among the psychometricians is evidenced by Vernon's (1979, p. 232) somewhat plaintive declaration that "*g* is not merely a cultural 'invention' of Western civilization." More and more people are beginning to wonder whether this isn't precisely what it is.

I.Q. data show that for middle-class British and Americans a figure measured on a quantitative scale correlates to a certain degree between parent and offspring and is predictive of achievement in the educational system and of the level of socioeconomic status attained. When the tests for obtaining this figure are given to groups in other classes and cultures, the results are far less dependable in predicting educational achievement or the status attained in the different cultural setting. But peoples of widely different cultures, who generally perform badly on American intelligence tests, nevertheless function with complete competence in their own milieu. In the harsh environment of the Kalahari Desert where a Westerner could not survive, the Bushmen feed, shelter, and protect themselves by using the complex body of assimilated and integrated information generated by their culture as a basis for making judgments (Howells 1954; see also the review by C. Shire of *The !Kung San* by R. B. Lee in *Science* 210:890 for later references on the Bushmen of which the !Kung San are one group). Among the people of Puluwat Atoll in the Caroline Islands, highest status is granted to the navigators who have mastered a locally developed system of seamanship that enables them to sail outrigger canoes over courses of from sixty to eighty miles of open ocean where the target island is only a mile or two in diameter. The people of Puluwat, even the master navigators, don't do very well on Western-designed psychological tests. Those who do best are the ones who have studied at a school of Western type on a neighboring island (Gladwin 1970). Tests could probably be devised which would classify these people within their own cultures in a way predictive of performance and achievement, and there would almost certainly be correlations between parent and offspring.

In writing of the children of the Guatemalan highlands, Kagan observes:

"The San Marcos child knows much less than the American about planes, computers, cars, and the many hundreds of other phenomena that are familiar to the Western youngster, and he is a little slower in developing some of the basic cognitive competences of our species. But neither appreciation of these events nor the earlier cognitive maturation is necessary for a successful journey to adulthood in San Marcos. The American child knows far less about how to make canoes, rope, tortillas or how to burn an old milpa in preparation for June planting. Each knows what is necessary, each assimilates the cognitive conflicts that are presented to him, and each seems to have the potential to display more talent than his environment demands of him. There are few dumb children in the world if one classifies them from the perspective of the community of adaptation, but millions of dumb children if one classifies them from the perspective of another society" (1973).

In view of these facts and because heritability is a characteristic of a population in an environment, one is justified in being skeptical of a fixed value for the heritability of the I.Q. The lack of clarity as to exactly what the tests measure, the failure of the tests to take account of cultural influences, and our complete ignorance of the nature of the environmental ones that are effective, make it questionable whether the studies that have been made tell us anything precise about the heritability of the I.Q. They rule out almost no hypothesis. Jensen (1969a) and Jinks and Fulker (1970) accept the 0.8 figure; Jencks (1972) thinks it should be something less than 0.5; Kamin (1974) sees no reason to think it differs from zero. Layzer (1974) and Feldman and Lewontin (1975) question whether measuring the heritability is conceptually valid at all. From a common sense point of view it seems reasonable to assume a genetic component in cognitive behavior. It requires a brain and certainly genes contribute to the structure and functioning of the brain even though we have no inkling of just how they contribute to cognition. But when one reads the studies and becomes aware of the continuously unsuccessful struggle to collect accurate and comparable empirical data, one cannot help wondering whether the claims to precision of the estimates turned out by the kind of ponderous statistical machinery at work in Jinks and Fulker's 1970 paper are not spurious. Heritability studies in domestic or experimental laboratory animals or plants where the goals are selective breeding or the understanding of evolutionary change are one thing. Such studies of human intelligence, even if they could be done, do not further such goals. Instead, they reinforce existing prejudices and intensify suspicion and misunderstandings. I.Q., achievement, and performance tests may certainly, if judiciously used, be of practical value in screening or classifying individuals for educational or vocational purposes. In the present state of our knowledge we would do better to suspend judgment on the inheritance of the scores.

The relevance of the heritability of intelligence to the subject of race is to be found in its most crucial phase in the widely held contention that American blacks are genetically inferior to whites in intelligence. The argument has been made in its most presentable form by Jensen (1969*a*, pp. 81−82). He points out that the mean of the I.Q. scores of American blacks is one standard deviation below the mean of the whites. Then taking the figure 0.8 as the proportion of the variance of the I.Q. distribution contributed by genetic factors, he argues that the elimination of all environmental influences working to reduce the black scores could not possibly raise the mean for the blacks by one standard deviation. One could put the argument in numerical form as follows. The variance of the I.Q. distribution is 225 (15^2); 20% of this is attributable to environmental influences. If 225 is reduced by 20%, we get 180 as the new variance corresponding to a new standard deviation of 13.4 ($\sqrt{180}$). This is a difference of only 1.6 from the present standard deviation. Hence if all environmental factors worked to reduce black I.Q.s and if these factors were eliminated for the blacks alone, this could not possibly raise the black mean by the 15 points necessary to bring it up to the value of the white mean.

There are many flaws in this argument, but the most glaring is the assumption that a figure for heritability based on data collected in one population is applicable to another. Jensen himself makes the point (1969*a*, pp. 64−65) that the major studies of the heritability of the I.Q. have been made on samples from the white population of Europe or the United States and that there are no adequate studies based on data from the Negro population of the United States. The 0.8:0.2 apportionment of the variance between genetic and environmental factors does not mean that such an apportionment would be characteristic of the white population under other environmental conditions. It has no relevance whatever to another population living in an entirely different environment.

What makes the problem almost inaccessible to scientific investigation is that to a very considerable extent the difference in environment between blacks and whites in the United States is determined by whether the individual is classified as black or white. In such a situation randomizing the environment is an impossibility. The fact is that we have no evidence on the question of the inheritance of differences in intelligence between races and we are not likely to get any until we discover means for greatly improving our techniques of investigation.

Stature, like intelligence, appears to be controlled genetically by a polygenic system, and Jensen (1969*a*, p. 60) applies his conception of a fixed upper limit to this character, too. Nutrition, he argues, is to stature what environment is to intelligence: if nutrition is grossly deficient, growth will be stunted, but above a minimal level of adequacy, dietary variation will have little effect on stature and heredity will be the primary cause of individual differences in height. In the studies of monozygotic twins raised apart, the correlations between twins for stature are consistently higher than those for I.Q. Consequently, one would expect a general increase in stature for a population as a result of environmental change to be even less likely than a rise in the average I.Q. Yet, for the past several decades, the mean stature in western Europe, the United States, and Japan has been increasing consistently and substantially. In the United States from 1918 to 1958 the mean stature of adult males rose from 67.03 inches to 68.20 inches, an increase of 1.17 inches, or 0.46 of one standard deviation (U. S. Public Health Service 1965).

The summer 1969 issue of the *Harvard Educational Review* carried a reply by Jensen (1969*b*, p. 229) to critics of his earlier article. In this reply Jensen endeavors to explain away the data on increase in stature as largely the result of genetic factors. These genetic factors he sees as changes in the mating patterns which have been taking place over the past century. With the increase in mobility, marriages within a single village or limited neighborhood have become less frequent and those between persons of more distant origins more so. This decreases the likelihood that individual alleles will be homozygous. There is much evidence from experimental animals that increased heterozygosity is accompanied by more rapid growth and larger ultimate size, this effect being known as *heterosis*. But the demonstration of heterosis in human populations is difficult. No human populations are inbred to the same degree as experimental animals; and the period during which the distance between the points of origin of spouses has been increasing has also been characterized by steady improvements in nutrition. Some studies have shown that the offspring of parents who are cousins or who come from the same village are shorter than the general population. But in any large human population the percentage of cousin marriages is rarely more than 2 or 3 percent and commonly even less, and decreasing the degree of inbreeding in that portion of the population more distantly related than, say, second cousins is not likely to produce detectable heterosis. In contrast, where increase in height has been looked for in interracial crosses, it has not been found (Morton 1962). In addition, the steady increase in the height of twenty-year-old Japanese males, which has been going on since 1900, was interrupted during World War II when nutritional levels were severely depressed (Kimara 1967). To conclude that the general increase in human stature was almost entirely attributable to genetic factors, one would have to contort the available evidence. Since stature showed so high a heritability in the twin studies and yet has shown so pronounced an increase over the past century, there is absolutely no reason to believe that a similar increase in the I.Q. is any less likely to occur.

SOCIOBIOLOGY

Although the psychological hereditarians presently seem to be suffering a partial eclipse, genetic determinism of behavior has recently found a new road to popularity in the doctrine of sociobiology put forth in 1975 by E. O. Wilson in a 697-page volume entitled *Sociobiology: The New Synthesis*. The central theme of the theory is that since all life is the result of evolution effected by natural selection, the behavior of a species, as well as its anatomy, is adaptive and exists because those who performed it have left more offspring than those who did not. Hence it has a genetic basis. As thus stated, there is little difference between this idea and the "species specific" behavior of Lorenz and the other ethologists. But sociobiology is a broader formulation that adds several other theoretical concepts, the most important of which are *kin selection* and *inclusive fitness*.

Since the time of Darwin there has been a problem of explaining why natural selection should not rub out altruistic behavior wherever it occurs because the altruistic benefactor should, in general, leave fewer offspring than the selfish beneficiary. W. D. Hamilton (1964) proposed an elaborately worked out solution to this problem, contending that if the altruist benefited primarily close kin, his genes (which presumably produce the behavior) will increase relative to those of the egoist. He contended that this process explained the development of the intense cooperation of the social hymenoptera (bees and ants) where sterile daughters of the queen spend their lives caring for their younger sisters instead of producing offspring of their own. The hymenoptera have a unique form of chromosome transmission which results in sisters having three-fourths of their genes in common but sharing only half their genes with their daughters if they were to have any. This extreme form of kin selection, he argued, has led to extreme forms of altruism. In birds and mammals, who lack the special hymenopteran chromosomal impetus to kin selection, cooperative behavior has developed because life in small groups of close relatives has facilitated it.

A corollary of kin selection is the concept of inclusive fitness. The fittest individual is not merely the one who leaves the largest number of offspring, but the one who acts, both before and after producing them, in a way that will maximize the probability that his genes will be transmitted to still later generations. This idea was further developed by R. L. Trivers (1971) who suggested that *reciprocal altruism*, aid rendered to a nonrelative when the chances of reciprocation are good, could enhance one's inclusive fitness. He also considered (1972) what degree of *parental investment* in a given offspring would contribute most to the inclusive fitness of the parent and later (1974), how the drive of both parent and offspring, each to maximize its own inclusive fitness, would lead to conflict between them.

By seeking the origin of behavior in selection, sociobiology emphasizes genetic determinism, conflict, and violence. It recognizes that many traits result from polygenic interactions, but tends to think of evolutionary change as resulting from single gene mutation. The extent to which sociobiology plays down the influence of environment and experience in the development of individual behavior is shown by Wilson's quoting with approval (1975, p. 560) Robin Fox's (1971, p. 284) statement that if children raised in isolation, as in the reputed experiments of the Pharoah Psammeticos and James IV of Scotland, had survived, "I do not doubt they could speak and that, theoretically, given time, they or their offspring would invent and develop a language despite their never having been taught one." This lack of understanding and appreciation of the role of culture and the way in which it interacts with the potentialities of the genotype during development has aroused severe criticism from the cultural anthropologists (Sahlins 1977, Montagu 1980).

In the writings of the sociobiologists there is to date little discussion of human races. The word is not in the index of Wilson's 1975 work. But the emphasis on genetic determinism of behavior is an invitation to explain group differences on the basis of different alleles. The psychometric hereditarians hypothesize a polygenic determinism for the cognitive limits of the individual. The sociobiologists go much further. All action of consequence in social or interpersonal behavior, since it is adaptive, has an important genetic basis. This includes practically everything from bonds formed between individuals, to sexual division of labor, to child rearing. "Human beings intuitively avoid incest . . . guided by an instinct based on genes" (Wilson 1978, p. 38). According to Wilson, the time has come "for ethics to be removed temporarily from the hands of philosophers and biologized. . . . The question that science is now in a position to answer is the very origin and meaning of human values, from which all ethical pronouncements and much political practice flow" (Wade 1976).

The methods for testing and establishing hypotheses based on this terrifying credo are not self-evident. The potentialities for using it to justify racial or class discrimination are infinite. Any deviation from "adaptive" behavior must result from bad genes whose carriers must obviously be treated without remorse to protect those lucky enough to have only good ones.

SUMMARY

Behavior in its broadest sense is as much a part of the phenotype as physical morphology. Both are intimately related to the genotype, but neither is rigidly determined by the genetic information. The morphology of the mature individual results from the interaction between his genetic information and the information in his environment during development. Morphology is innate in the sense that the genetic information directs its formation; but no element of form or structure is innate in the sense that its development is inevitable and unpreventable. The child born with cataracts because of a fetal rubella infection had the same innate potential for transparent lenses as the normal child, but it was not actualized. Similarly, all behavior is innate in the sense that if the genetic information that produced the individual did not exist there would be no individual to act; but no behavior is innate in the sense of being inevitable. Behavior patterns arise during development as a result of interactions between the information in the organism and that received from the environment. The environmental inputs constitute experience, and the interactions between behavior and experience are more prolonged, more complex, and more versatile than the individual-environmental interactions affecting morphology. Development and experience within the normal environmental range will produce individuals with behavior within the range recognized as species typical. From individual to individual there will be variability of behavior characteristic of the interaction of the array of genotypes in the population with the range of environmental influences. In human populations, particularly with respect to behavior, class or status variability must be recognized in addition to individual and geographic variability. Behind all this variability, making it possible, is a complex of interacting polygenic systems. Thinking in terms of unit characters and innate versus learned behavior is inappropriate to understanding the complexity of biological reality.

5

THE ORIGIN AND ROLE OF VARIATION

RECOMBINATION

The origin of the genetic contribution to the variation between individuals is of two sorts: recombination and mutation. In the higher animals every individual is diploid: he carries two versions of the genetic information of his species, one derived from his mother and the other from his father. Together, these two versions constitute his unique genotype, one genetic complement. When the mature individual forms sperm or eggs, the genetic material in these sex cells, or gametes, is half the diploid amount, just one version of the genetic information. In man one version consists of 23 packages, or portions, called chromosomes. The diploid cells of the body thus have 46 chromosomes, 23 of maternal origin and 23 of paternal. The 23 chromosomes of each version are all different from each other, but every one has a similar partner, or homologue, in the other version. Thus there are 23 pairs of homologous chromosomes. When a sex cell is formed, the single member of each individual pair is drawn at random from that pair in the cell of the gonad that produced it. So one sperm, although it contains half the genetic material of the individual producing it, may have from zero to 23 of the chromosomes that that individual received from his mother. As a result, a sperm produced by an individual may carry at two different loci a combination of alleles that occurred in neither the sperm nor the egg which joined to produce him. When such a sperm fertilizes an egg, the chromosomal complement of which has been similarly formed, the resulting zygote may, as a result of interaction between the alleles at the two loci, produce a phenotype different from either parent or perhaps even unknown in either parental line for several generations.

It is not easy to find a clear illustration of this process in human populations because we cannot make the crosses necessary for a demonstration, but many such cases are known in experimental animals. For example, there are strains of chickens with pea combs and others with rose combs (fig. 8). Both strains breed true, but when the two are crossed, all the offspring have a third type of comb, called walnut. If these walnut phenotypes are interbred, among their offspring

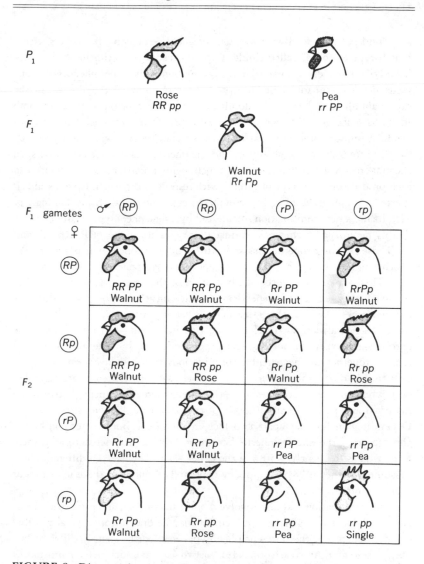

FIGURE 8. Diagram showing the interaction of alleles for comb form in domestic chickens. A cross between a hen from a true-breeding rose comb line and a cock from a pea comb line gives offspring all having walnut combs. These F_1 birds all produce four kinds of gametes which can come together in the sixteen combinations for which the genotypes and phenotypes are shown in the checkerboard. Recombination of the alleles produces in the F_2 not only the walnut comb of the F_1 birds but also the rose and pea combs of the original parents(P_1) and a fourth form of comb (single) not manifested in either parental line. (After Sinnott, Dunn, and Dobzhansky 1958, *Principles of Genetics*, McGraw-Hill.)

are found not only fowl with walnut, rose, and pea combs, but others with a fourth type of comb, called single. Careful breeding experiments have established that two loci are involved in the production of these four phenotypes. One locus has a dominant allele that produces rose comb in the absence of the dominant allele at the other. A dominant at this other locus produces pea comb in the absence of a dominant at the rose locus. Dominant alleles at both loci produce walnut comb, and two recessives at both loci produce single comb.

Figure 6 shows the genotypes and the phenotypes in these two crosses. In the first cross, the rose hen and the pea cock, being homozygous at both loci, can each produce only one type of gamete with respect to these loci: Rp eggs and rP sperm. A union of these two gametes produces a genotype with one dominant at each locus, a new combination manifested by a new phenotype, walnut. These doubly heterozygous chickens produce four types of gametes: RP, Rp, rP, and rp. These can come together in sixteen different ways, one of which gives rise to another new combination, pprr, the genotype corresponding to single comb.

This is a simple, textbook illustration of the process. Only two loci are involved; they segregate independently, which means that they are on different chromosomes; there are sharp distinctions between the phenotypes. Recombination is not always so simple. If the two interacting loci are on the same chromosome, they do not segregate independently, but the alleles tend to remain together in the parental combinations rather than appearing in the offspring in new combinations. Alleles on the same chromosome are separated only when another process occurs—the interchange of segments between the maternal and paternal chromosomes of the same pair, known as *crossing over*. Depending on how far apart two loci are along the linear extension of the chromosome, the alleles inherited from one parent may separate at gamete formation as frequently—50% of the time—as if they were on different chromosomes, or they may part company very rarely—once in a thousand times or even less often.

Recombination may also involve more than two loci, and the two alleles at any given locus in a genetic complement may be drawn from several possible choices. Then too, the different phenotypes may not be sharply different but may instead grade into each other. It is easy to see that under such circumstances genetic analysis of the process becomes much more difficult and complicated. This leads to exactly the situation described in chapter 2 as a polygenic system.

The number of possible genotypes in a polygenic system goes up very rapidly as the number of loci and the number of alleles at each locus increases. With only four loci and two equally frequent alleles at each locus there will be $81(3^4)$ different genotypes, and the rarest of these will occur only once among $256 (4^4)$ zygotes. As we increase either the number of loci or the number of alleles the number of genotypes multiplies very rapidly and the probability of finding the rarest genotype goes down. If the different alleles are present, as is almost always the case, in unequal frequencies, this makes the rarest genotype even rarer. The result is that in higher animals the number of possible genotypes in a complex polygenic system may easily be greater than the total number of individuals in the entire population of a species. Some genotypes may occur only once in several generations, and other potential ones may never occur at all.

Recombination is thus a creative force in biology, producing rare or completely new genotypes that can be tested in the environments where they occur. The production of new zygotes in every generation is a chance process, which may at any time produce something entirely novel. But the precise results, while in accordance with the laws of probability, are unpredictable with respect to any individual zygote.

MUTATION

Recombination brings different alleles together. If every locus had only one genetic message, there could be no recombination. The existence of genetic variability in the form of distinct alleles is necessary for the process. The differences between alleles are the result of mutation. In its simplest form, mutation is the substitution in the DNA of one nucleotide pair for another. Such a simple substitution may cause one amino acid to be replaced by another in the corresponding polypeptide. To change the glutamic acid in the sixth position in the beta chain of human hemoglobin to valine—and thus produce hemoglobin S—the thymine in the center of the triplet need only be replaced by adenine, making the corresponding change in the transcribed RNA from adenine (A) to uracil (U). A single substitution may also cause a polypeptide to be terminated prematurely. This is very likely to produce a completely nonfunctional protein.

Single substitutions are included in what are called *point mutations*, a term that goes back to the days of premolecular genetics. Such mutations were given this name because they appeared to be within the functional gene and there was no cytological evidence of change in chromosome structure. There are other types of point mutations besides single substitutions; many of these are well known in microorganisms where they have been extensively analyzed; our direct knowledge of them in human genes is only beginning to accumulate.

The chemical reactions that produce point mutations result from several causes. Accidents occur with low frequency but continually in the everyday activities of the atoms and molecules of the genetic material, and these account for some mutations. Certain chemicals induce changes and are thus mutagenic agents. Ionizing radiation—naturally occurring in the form of cosmic rays or coming from radioactive isotopes or man-made radiation such as X rays—also produces mutations.

The variability existing in any population at any given time is the result of previous mutation. Every population always has a reservoir of variability and its nature is determined by what sorts of variants mutation has been putting into the population and which ones natural selection has been throwing out. The process is continuous and any mutation that has a finite probability of occurring has probably already occurred and is very likely lurking somewhere—perhaps in a heterozygote as an unexpressed recessive, in an extremely rare phenotype, or in a genotype in which alleles at other loci suppress its potential phenotypic effect. Populations do not mark time waiting for a certain mutation to occur. They make do with the variability they have.

Both mutation and recombination are random in the sense that one cannot predict individual events. In any given population, however, different mutations have different rates of occurrence, and different types of recombination occur with different frequencies. Both mutation and recombination are very probably influenced by the overall integration of the total genetic program; and, as a part of this program, they must be influenced by natural selection. In the genus *Drosophila*, the vinegar flies, numerous mutations affecting the pigments that color the eyes keep recurring, each with quite predictable frequency. *Drosophila* eyes are generally reddish, varying from species to species from bright scarlet to deep maroon. Within the species, mutants run this whole gamut and extend to white at one extreme and to dark brown at the other. But no mutation has ever been observed which produced a fly with green eyes. In other families of flies, however, green eyes are often the norm.

Although all the genetic variability in a population has arisen from past mutation, the amount of mutation in any one generation is almost always very small compared to the accumulated reservoir. Mutation rates for given loci vary, but they are commonly of the order of 1 in 10,000 (10^{-4}) or 1 in 100,000 (10^{-5}) gametes per generation. But unless they are dominant, even deleterious mutants are not eliminated at once from the gene pool. A recessive lethal accumulates until its frequency is the square root of its mutation rate. A recessive lethal occurring 10^{-4} per generation would at equilibrium have a frequency of 10^{-2}, or 1 percent, even though the increment per generation would be only 0.01 percent. In this case the reservoir would be one hundred times the increment per generation. For less deleterious mutants the disparity between increment and reservoir level is much greater. Consequently, the ability of a population to respond to selection is not likely to be affected much by its present mutation rate. Doubling or tripling a mutation rate will not have much effect on gene frequencies until after many generations.

When two dissimilar strains of animals are crossed, phenotypes that were unknown or extremely rare in the parental strains sometimes turn up in the second or third generations. In such cases it is often very difficult to decide whether one is dealing with the results of unusual recombination or an increase in the mutation rate.

In all populations of animals there are certain defective alleles—most of them recessive—which show up as obviously ill-adapted phenotypes when two such alleles for the same locus occur in the same individual. Human populations carry such defective alleles. Phenylketonuria, galactosemia, and cystic fibrosis are all examples. They are relatively rare; cystic fibrosis, the most frequent of these three, occurs among Europeans once in about 2,500 live births. The affected persons are severely disadvantaged and rarely reproduce. The frequencies of these genetic diseases vary to a considerable extent from one human population to another, but grossly defective alleles producing pathological conditions when homozygous constitute a very small part of the genetic variability that enables a population to respond to an environmental challenge. The reservoir of variability on which a population draws to shift its modal phenotype is made up primarily of the slight differences among the normal alleles, the *isoalleles*. When a population of experimental or domestic animals is subjected to artificial selection, it is possible within a few generations to cause a shift either up or down in the mean of almost any character that can be accurately measured. Such a shift is not usually accompanied by any substantial increase in the proportion of grossly abnormal individuals. The change results from reshuffling the different normal alleles, not from increasing the frequencies of the defective ones.

In addition to point mutations—changes in the information within the functional genetic units—there are also larger-scale alterations of the genetic material. These include deletions, duplications, and various kinds of rearrangements of the genetic units within the chromosomes. They are often referred to as chromosomal changes or chromosomal aberrations. This kind of genetic variation is known in many kinds of animals, and in some cases chromosomal differences characterize different races within the same species. Chromosomal aberrations are known in the human species, but they are rare and accidental; and there is no evidence of any differences between human populations in chromosome structure or morphology.

Probably the best known case of a chromosomal aberration in man is what is known as Mongolism, Down's syndrome, or trisomy 21. This is a very characteristic abnormality recognizable at birth by physical stigmata. As the affected child develops, he shows severe mental retardation. The condition was first recognized in 1866 by a British physician, A. Langdon Down. He called it the Mongolian type of idiocy because one of the diagnostic characters—a fold of skin over the inner corner of the eye—suggested to him a connection between the abnormality and East Asian populations. That Down considered this more than an accidental similarity is shown by his mention in all seriousness of finding also among his idiot patients Ethiopian, Malay, and American Indian types.

In 1959, Lejeune, Gautier, and Turpin demonstrated that Down's syndrome was the result of an unbalanced genetic complement in which there are three number 21 chromosomes instead of the usual pair (hence the alternate name, trisomy 21). Affected individuals have, therefore, 47 rather than 46 chromosomes. It is known that the condition comes about by an aberrant division in the maturation of the egg. Both number 21 chromosomes, instead of just one member of the pair, go into the haploid nucleus. This abnormal kind of division occurs with a certain low frequency, and the probability of its occurring in any individual female increases with her age. The abnormal division can occur in the formation of a sperm, although this seems to happen much less often. The aberration is not confined to any race or population and, of course, it has no more connection with Asiatic populations than with European.

Some cases are known where a person with the typical characters of trisomy 21 has only 46 chromosomes. The third chromosome 21 in these cases is fused, or translocated, to another chromosome. In a few such cases the mother of the affected individual has been found to have only 45 chromosomes, the fused chromosome representing both chromosome 21 and some other chromosome, most likely number 13, 14, or 15. This parent is phenotypically normal because she has one complete, balanced genetic complement. In the egg that produced the affected offspring the mother's fused chromosome and her single chromosome 21 both went into the haploid nucleus. When this was fertilized by a sperm containing one chromosome 21, the resulting zygote had three chromosomes 21.

In general chromosomal changes or rearrangements produce abnormal individuals only when they result in unbalanced genotypes with more or less than one complete diploid set of genetic information. Some rearrangements, even when they occur in balanced form, reduce the fertility of the otherwise normal individual by causing the production of some unbalanced gametes. Chromosomal rearrangements occur in human populations, but neither differences in types nor in frequencies of rearrangements appear to explain or to be significantly related to the phenotypic differences between human races.

NATURAL SELECTION

Since every population is genetically variable, and since in every generation this variability is recombined with some degree of randomness, there is in every generation an array of different genotypes, most of them approximating the modal phenotype, but each having slightly different potentialities for coping with the vicissitudes of the environment. Because of these different potentialities some genotypes will produce more offspring, others fewer. If a population of constant size has been breeding at random for many generations in a given locale with a substantially stable environment, and if the number of immigrants and emigrants is small, the population will probably be close to equilibrium. This means that the array of genotypes will remain approximately the same from generation to generation. New genotypes will be those that, with the allelic frequencies obtaining, are so rare that they occur less frequently than once per generation. Alleles that find themselves more consistently in genotypes with less-than-average reproductive success will be reduced in frequency; those that are more often than not in genotypes having more than the average number of offspring will become more numerous. At equilibrium the frequency of every allele will be theoretically constant, for the number eliminated by reproductive deficiency or by mutation from the allele will be equal to the number brought into the population by mutation to the allele.

Natural selection is the process by which differential reproduction eliminates certain alleles from the population and replaces them with others. It does this not by the creation of new alleles, but by the differential reproduction of those already in existence. When a population is at equilibrium, selection keeps the allelic frequencies constant. As a result, the array of genotypes remains constant and the modal phenotype does not change. The process of maintaining this equilibrium is called *normalizing selection*. For most quantitative characters, maximum fitness—that is, the highest probability of leaving not less than the average number of offspring—is associated with the mean value. Individuals at the extremes of the distribution are likely to leave fewer offspring, on the average, than those found near the middle. So, at equilibrium, under normalizing selection, in every generation the tails of the distribution are lopped off; and in the next generation these tails are regenerated by recombination. To illustrate, let us reexamine mating 2 in table 2. Even if we assume that genotypes with four minus alleles are sterile, we see that they will not disappear from the population, because in every generation some of them will be produced by recombination in matings where both parents have at least one minus allele at all four loci involved. If we think not of a single character, but of all characters combined, we can picture, in place of a single normal distribution in one plane, a solid dome produced by rotating a normal curve around its mean. The infinity of characters would be represented by diameters through the perpendicular at the common mean. Normalizing selection keeps trimming off the thin edge around the bottom of this dome.

Directing selection, as distinct from normalizing selection, is the process by which the mean of a character is changed. Programs of artificial selection are of this nature. A sheep breeder interested in increasing wool production will select the parents in each generation from those sheep that produce more than the average weight of wool at shearing. The result will be an increase from generation to generation in the mean weight of wool produced, the amount depending on the heritability of wool production in the flock. In a given period under natural conditions in a wild population an increase in the mean of some character may add to the fitness of the animals. Then natural selection becomes directing. During the Miocene and Pliocene periods, when large areas of grasslands appeared and spread, some species of horses became grazing animals. Grass contains silica and is extremely abrasive to teeth. Horses with low teeth soon wore them down and were at a disadvantage. During these two periods the height of horses teeth increased manyfold, an excellent example of long-continued directing selection.

Studies of artificial directing selection have uncovered several interesting and important characteristics of the process. The mean of almost any character in sexually reproducing animals may be changed by selection, but the rate and degree of response vary enormously. Response to directing selection is usually accompanied by changes other than the one selected for; and as the mean of the selected character is pushed farther and farther from its original position, the population declines in general fitness and becomes harder to maintain. If, in such a population, selection is discontinued—mating is allowed to occur at random rather than on the basis of the selected character—the mean of the selected character will, in the next few generations, drift back toward its original value; and as it does so, the overall fitness of the population will recover. This is a very general phenomenon and a very important one, well known to animal breeders and to students of selection. Michael Lerner (1954), who has done more work on this problem than anyone else, has called this process *genetic homeostasis*.

That a population at or near equilibrium tends to be genetically homeostatic means that the pattern of allelic frequencies is not accidental but constitutes an integrated system related to gene interaction during development, to the means of the many different characters in the modal phenotype, to the total array of phenotypes in the population, and to their relative fitness in a given environment. There is also impressive evidence that the pattern of allelic frequencies characteristic of a population is maintained because often a heterozygous combination of two different alleles has a selective advantage over either allele in homozygous condition. Any tampering with the allelic frequencies will change not just one character but also several others. To produce a lasting change in the mean of a given character, selection must be continued until compensatory changes in several characters and in many allelic frequencies have produced a new stable, integrated, equilibrium. Selection can then return to a normalizing rather than a directing role.

Normalizing and directing selection can and do go on in the same population at the same time. No population ever reaches complete equilibrium.

There are always fluctuations from generation to generation in environmental pressures and in population response. In natural populations, directing selection generally acts at a mild level of intensity in a background of normalizing pressure. This allows directed changes in characters to be integrated into the overall biotic program as they slowly occur. During the Cenozoic period, horses' teeth increased in height by approximately sixty millimeters, but this occurred over a period of sixty million years. Although the increase was not uniform throughout the period, it averaged about one millimeter per million years, a rate that could be reconciled with simultaneous adjustments of the whole program of the species to the change.

Populations are always probing the environment with new behaviors; and from time to time a new way of doing things that increases fitness is discovered. Then directing selection goes to work to increase the efficiency of the new technique. During the same Cenozoic period the ancestors of the giraffes found a food source in the leaves of trees. To exploit this new method of feeding, a long neck proved advantageous; and directing selection very effectively perfected one.

Natural selection is often popularly pictured as a process of culling. Merciless elimination by slaughter purges the population of its defectives and leaves only the strong and fit. There are species of animals, of course, in which the offspring in every generation are many times the number that can possibly reach reproductive age. The result is enormous mortality. In other species, however, the number of offspring is low and so is the mortality before adulthood. Natural selection acts in both cases. As long as there is differential reproduction of different genotypes, changes in allelic frequencies will occur. If genotype A leaves four offspring that survive to reproduce and genotype B leaves only two, the alleles found in A will be more numerous in the next generation than they were in the last with respect to those found in B. This will be true no matter what the mortality.

Part of the effect of normalizing selection is to discard the grossly deleterious recessive genes when they turn up in homozygous condition. The alleles that produce galactosemia and phenylketonuria are held by this process at a frequency level determined by the mutation rate and by the probability that their homozygous carriers will not reproduce. But this type of gene elimination is only a small part of the effect of normalizing selection. The picture of selection as a stern judge constantly eliminating bad guys and rewarding good guys among the genes is a distorted one. Selection gets its results by trying out new combinations among the dependable run-of-the-mill alleles. Different combinations of alleles can produce the same phenotype. One population will use a certain group of combinations, another will use another group, or two populations may use overlapping groups.

In the experimental program for developing populations of *Drosophila* resistant to DDT described in chapter 2, several different populations were subjected to selection separately at the same time. It was possible to show that the three of these which developed high resistance did so by different genetic means. In all three polygenic systems were responsible for the resistant phenotype; but the systems differed in the distribution of the factor among the chromosomes and in whether the effects of the factors were merely additive or had dominant interactions.

The ABO blood groups in man provide an illustration of selection for combinations of alleles. Anchored in the membrane surrounding the hemoglobin-containing cells of the blood are projecting chains made up of sugars. In many individuals these chains end with a galactose in the penultimate position and a fucose at the end. Persons having only such chains are said to be of blood group O. Some other individuals have an enzyme that adds another molecule of galactose to the chain, fastening it to the galactose in the penultimate position, so that the end of the chain is forked. These individuals are of blood group B. Still other individuals have a different enzyme that adds another sugar to the chain at the same position, but instead of adding a simple galactose it adds a modified galactose known as N-acetyl-galactosamine. These individuals have blood group A.

The inheritance of these three types of endings on the polysaccharide chains is such that we can be sure it is controlled by one locus that can be occupied by any one of three different alleles. One of these, which we call O, leaves the galactose-fucose end unaltered; another, B, produces the enzyme that puts on the second galactose; the third, A, makes the enzyme that adds the N-acetylgalactosamine. Since every individual has two of these alleles, every person has one of the following combinations: OO, OA, AA, OB, BB, and AB. The first is phenotypically O, the second and third A, the fourth and fifth B, and the sixth AB. Only in individuals of the OO genotype are there no other chains than those with the O-type end. In all other genotypes there will be either an A enzyme or a B enzyme or both, and as a result, many of the chains will have been supplemented with an additional sugar. In the AB genotype both enzymes are present; so some chains are supplemented in the B manner and others in the A.

These chain endings are antigens against which antibodies are produced, except that no one produces antibodies against his own chains. So the O phenotype has anti-A and anti-B antibodies; the A phenotype has anti-B antibodies; the B phenotype has anti-A antibodies; and the AB phenotype has neither. These antigen-antibody relationships complicate the giving of transfusions, but what other function these chains perform is a total mystery. None of these alleles has any demonstrable superiority over the others, but populations in which only one allele is present are extremely rare. Figure 9 shows the

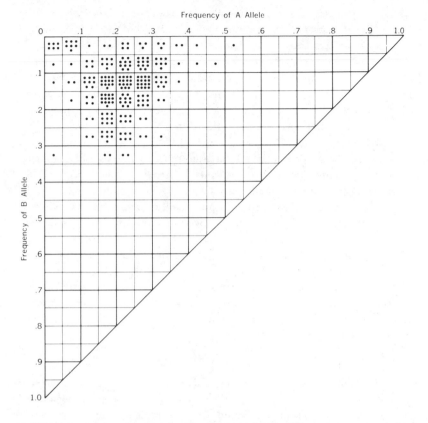

FIGURE 9. A plot showing the distribution of 215 different human populations classified on the basis of frequencies of blood group alleles of the ABO system. If the relative frequencies of these alleles were determined by chance, the entries in this plot would be scattered at random over the entire triangle. This is clearly not the case. (After Brues, 1954.)

distribution of 215 human populations classified according to frequencies of these ABO alleles. Along the upper leg of the triangle are the populations with from 0 to 100 percent of allele A. Along the left-hand leg are the populations with from 0 to 100 percent of allele B. Along the hypotenuse would be those populations with 0 percent allele O.

The striking thing about this figure is that the entries are crowded together in a restricted part of the triangle. If the ABO alleles were equal in adaptive value and there were no adaptive interactions between them, we would expect the populations to be scattered at random over the whole triangle, for the relative frequencies in different populations would then be determined by chance. But the observed distribution makes it clear that this is not so. There appears to be an optimal set of frequencies where the populations tend to cluster. This lies between 15 and 30 percent for A and between 5 and 20 percent for B, allowing O to vary from 50 to 80 percent. It is hard to see how these relative frequencies are maintained unless some combinations of alleles have greater adaptive value than others. It is quite clear that selection does not rank these alleles on a single scale of adaptive excellence and reject the two of inferior rank. The ABO system is genetically very simple. It includes only three alleles at a single locus. But even at this simple level one can get a glimpse of the working of the processes in polygenic systems which lead to genetic homeostasis. Most polygenic systems must contain several loci and many more alleles, and the interactions must be correspondingly more complex. In fact, the ABO system is a part of a larger polygenic system, for we know that there are interactions between the ABO alleles and alleles at other loci which influence red cell antigens. Until we have a better understanding of the functions performed by these antigens, it is unlikely that we will be able to analyze in a satisfactory way the manner in which selection maintains certain frequencies of the ABO alleles. But the ABO system shows us clearly that selection limits itself neither to casting out defective alleles nor to the maintenance of some favored allele at a frequency of 100 percent.

THE EBB AND FLOW OF VARIATION

Mutation is constantly pumping variability into the gene pool, where it accumulates. This variability is shuffled and redealt in every generation to produce the array of phenotypes found in the population. Natural selection— the result of differential reproduction of different genotypes—tends to keep the gene pool near an equilibrium at which the great majority of individuals is composed, generation after generation, of genotypes well adapted to the environment. This great majority is characterized by both genetic and phenotypic variability, but its members approximate the modal phenotype of the population.

It is obvious from what we know of mutation, recombination, and selection that for sexually reproducing creatures, a pure race or a genetically homogeneous population is a totally imaginary construction. Whatever else may be said for or against it, sex has the property not of destroying such genetic homogeneity, but of preventing its ever coming into existence. Two populations may differ with respect to one or more clearly observable characters, which they may possess in common but with different frequencies. But a careful search for subtler character differences will always uncover them, and any investigation of physiological and biochemical differences will surely turn up enormous variety. No two individuals are identical; no two populations can possibly be.

Whether a population is variable or uniform is relative and determined to a great extent by the kind of examination we give it.

If two populations that have been separated for many generations are brought together, and matings occur between individuals from the two, the offspring will show some new combinations of characters. Some of these novel phenotypes will appear in the first generation, others in the second, third, or even later. If mating within the merged population is at random, within a dozen generations or so the group will have come close to a new equilibrium in which the array of individual variants will be substantially the same from generation to generation. New combinations of characters will be rare. The two populations will have become one, and it may or may not be more obviously variable than the two that merged to produce it. All three populations, however, and all populations that ever have been or ever will be, are polymorphic. Polymorphism is an inherent property of living things.

No population, of course, ever reaches complete equilibrium. The environment is never constant, and in interacting with it a population is continually readjusting it genetic system. Few populations are completely closed, and those that are not are constantly receiving and assimilating immigrants who are never exactly equivalent to a random sample of the assimilating group. Even a closed population—on an isolated island, for example—unless it is very small and occupies a very small area, tends to break up into subpopulations. Influences such as distance, terrain, and habitat distinctions prevent completely random mating, and selection tends to operate slightly differently in the different areas and habitats.

Island populations often illustrate another process that influences divergence between different groups. Some years ago Mayr (1954) pointed out that on many of the smaller islands off the coast of New Guinea there are populations of birds of species found also on the mainland. Within a single species the island forms are often very different from the mainland form, and each island population has its own peculiar set of characters setting it off as a distinct subspecies.

The distinctness of these subspecies is without doubt attributable to what the statisticians call *sampling error*. If one reaches into a bag containing half beans and half peas that have been thoroughly mixed and takes out a handful, one will rarely get exactly the same number of beans and peas. The ratio will vary from handful to handful, sometimes in the direction of more beans, at other times of more peas. The larger the sample, the closer the sample ratio will be to the ratio in the bag. In the case of one very small sample, say six or eight seeds, the departure from the ratio in the bag may be very great. Island populations are very likely to have been started by a few founding colonists, perhaps a group as small as six or eight. Such a small number could not possibly reflect the actual allelic frequencies found in the large mainland population. If the founders flourish on the island and build a large population, its allelic frequencies will be greatly biased by the amount and kind of variability present in the original colonists and it may depart substantially from the allelic frequencies of the mainland. If two alleles both have relatively high frequencies on the mainland and confer a substantial increment of fitness when they occur together and if only one of the two happens to exist among the colonists of an island, that one may languish in the absence of its partner and drop to a low frequency as the island population expands and approaches equilibrium. It is, of course, very difficult to demonstrate rigorously that any natural population owes its distinctive character to the *founders' effect*, as this type of explanation is called, but most geneticists and students of evolution are convinced that the idea corresponds to a real and important process.

There is good evidence from laboratory experiments to support this hypothesis (Dobzhansky and Pavlovsky 1957). Two groups of twenty separate populations were set up with known frequencies of certain alleles. In one group, each population was started with 4,000 individuals; in the other, with only 20. After twenty generations the populations started with large numbers diverged much less from each other in allelic frequencies than did the populations started with only 20 founders.

The founders' principle is a special case of a more general phenomenon termed *genetic drift*. Whenever a gamete is formed, a sampling process occurs, for the gamete has only half the alleles of the diploid animal that produced it. Consequently, in a stationary population where, on the average, each member contributes only two gametes to the next generation, each genetic complement is sampled only twice, and a substantial portion of it will not be passed on at all. If the population is very large, the sampling error from generation to generation will not be great, for an allele not chosen from one genetic complement will be among the elect from another. But where populations are small, sampling error may cause substantial fluctuation in allelic frequencies, and it is certainly possible that this may sometimes shift the equilibrium of a polygenic system in one direction or another. Again, it is difficult to prove that genetic drift has occurred in natural populations, and undoubtedly the hypothesis has been invoked recklessly to explain all kinds of puzzling population differences. Nevertheless, the evidence that genetic drift can occur is good.

A study of a small human population in Pennsylvania made in the 1950s showed that for the M and N blood types the allelic frequencies of the generation under twenty-seven at that time differed significantly from those of the generation of their grandparents, aged fifty-six and over (Glass 1956). Two cycles of reproduction had been enough to shift the allelic frequencies substantially, apparently as a result of chance alone.

Every population has within it genetic elements that have come from other populations in the past, and some of its present genetic material will go to other populations in the future. At any given time one may map the subspecies and local populations within the subspecies and get a complicated set of intergradations. This is, however, a simplified view of the process going on through time during which the intergradations fluctuate, some populations separating while others merge and all interact continuously with their environments. But this pulsating, pullulating ebb and flow can take place only within the species, for it is based on successive shuffles and deals of the information coded in the genetic material. When the coded programs of two populations have diverged to such an extent that homologous sections of the genetic complement (which is analogous to the informational tape in a man-made cybernetic device) can no longer be interchanged without disrupting the biotic program—no matter at what point in the cycle the block occurs—they belong to different species. From this point on they can be to each other only part of the environment.

To go back to our juncos, we can see that the various subspecies have in the past been isolated from each other in varying degrees but must always have been able to exchange genetic material in order to do so now. From our knowledge of the present it is impossible to read all of the history of the species, but some guesses can be made. During the Pleistocene period the populations of the various subspecies probably expanded and contracted as the glaciers, both continental in the east and mountain in the west, covered more or less area and as differences in rainfall caused similar fluctuations in the size and location of forests. The great area from Nova Scotia to Alaska now occupied by the subspecies *hyemalis* suggests a recent expansion. It is probable that when the continental glacier extended as far south as Ohio and Pennsylvania, the juncos east of the Mississippi were isolated in the middle and southern Appalachians. Whether or not there was then a *carolinensis* population in the south and a *hyemalis* one in the north, with the retreat of the ice and the development of new habitable areas farther north and west, the eastern population must have spread northward to Alaska. The period of isolation from the western populations had not produced sufficient genetic divergence to produce two separate species, for whenever *hyemalis* has come in contact with western subspecies, there has been interbreeding—with *mearnsi, montanus, oreganus*. In fact the subspecies *cismontanus* appears to be a population of fairly recent origin composed of individuals whose ancestors were partly from *hyemalis* and partly from *montanus* populations. If we knew the complete history, we would probably find that for every one of the subspecies there was a period in the past during which it was compounded from two or more different populations.

In southern Arizona there is another population of juncos—*Junco phaeonotus palliatus*. These birds are somewhat similar in appearance to *Junco hyemalis dorsalis*, but the brown of the back is redder and the iris of the eye is yellow. This yellow-eyed junco has different habits, too. It spends more time on the ground and it walks or runs rather than hops. What is more, no evidence of interbreeding between *palliatus* and *dorsalis* has ever been found. Here speciation has occurred. *Palliatus* populations form a part of another complex of populations of Mexican juncos which intergrade among themselves, although they are all reproductively isolated from the populations to the north. At some earlier date the two species *Junco hyemalis* and *Junco phaeonotus* must have constituted a single species. *Dorsalis* and *palliatus* are too similar not to have come from a common ancestor. But in the course of time, isolation has produced such wide genetic divergence that the two groups of populations have become reproductively isolated and each is now evolving on its own.

It is a widely accepted idea that two species may be differentiated by some very simple genetic mechanism—different alleles at a single locus or at perhaps two or three loci. Carleton S. Coon (1962) described his idea of how an "old species evolves as a unit into a new species." This occurs, he says,

> When a genetically isolated population acquired a new and favorable hereditary trait that is controlled by a single gene or by a complex of genes operating in concert. Then the new trait gradually replaces the old one through natural selection (pp. 27−28).

Lorenz (1966) takes a similar position:

> What causes a species to disappear or become transformed into a different species is the profitable invention that falls by chance to one or a few of its members in the everlasting gamble of hereditary change. The descendants of these lucky ones gradually outstrip all others until the particular species consists only of individuals who possess the new invention (pp. 23−24).

This conception of speciation is so oversimplified that it is not merely misleading; it is false. In the "everlasting gamble of hereditary change," natural selection customarily operates with a small percentage in its favor rather than with a spectacular sure thing. So much delicate readjustment is constantly going on within the biotic program of a species that to picture that program being changed into another by the substitution of different alleles at one or a few loci is fantastic. Within the single species *Junco hyemalis* there is enormous variation not only in the color patterns of the feathers but in overall size; in relative length of wings, tail, or toes; in form and color of the bill—in fact, in almost any character that one examines. All these variations are possible within the single biotic program that characterizes the species. They can all be coded in the genetic material. There is corresponding variation within *Junco phaeonotus*. The programs of the two species, however, differ in a much more fundamental way than the local subspecific programs within each species differ from each other. For effective reproductive isolation to build up between two populations, their entire biotic programs are usually reconstructed by a long, slow accumulation of small increments of divergence. Most phenotypic differences between species are not produced by alternate alleles at single loci. Within each species each character is regulated by a polygenic system. The same character is not produced in two different species by the same constellation of alleles. Within a species the different populations achieve different equilibria by varying allelic frequencies at many loci. No one knows how many, but certainly dozens are involved, very likely hundreds. The very recent discoveries of intervening and other nontranslated sequences in the DNA and of the nuclear processing of RNA make clear how very large the number of alleles at each locus may be. The differences between species are both more extensive and more fundamental than those between subspecies. They are not achieved by the replacement of a few spare parts.

6

TRADITIONAL MISCONCEPTIONS ABOUT HUMAN VARIATION

A group of populations with individual variation within them and geographical variation between them, with varying amounts of interbreeding among them, each tending toward genetic equilibrium but fluctuating around it: this concept of the biological species is perhaps complicated, but there is no reason why it should be beyond human comprehension. Numerous geneticists and students of evolution—Dobzhansky, Mayr, and Simpson, among others—have pointed out that human races are the counterparts of subspecies in other animals, that there are no hard-and-fast boundaries between them, and that they differ not in kinds of alleles but in relative frequencies. But there are other concepts of race dating back to an earlier period which are still widely accepted even among the educated, and these very much becloud and distort the realistic appreciation of human variability.

The many attempts that have been made to give a systematic description of human races undoubtedly arose from the very human desire to bring order out of chaos, the same drive that led Adam to name the beasts and the animal taxonomists to wrestle with the species problem. But in dealing with human subspecies there have been peculiar complications. Man's interest in himself is naturally greater than his interest in other creatures, but this intensity of self-interest makes it difficult to be objective and by no means guarantees a self-understanding corresponding to the interest. A classification of races involves the position of one's own group and one's self, and one is loath to accept any system that threatens the psychological security of either.

The literature on human races is so vast that even a historical survey would require a large volume. Woven into much of the writing—even up to the present—are five complicating influences that continue to interfere with the recognition of biological reality: (1) the idea of the Platonic type, (2) a vision of pure races seen in a mirror reflecting the remote past, (3) the confusion arising from the similarity of race and social class, (4) the emotional overtones always present when one contemplates one's fellow men (and women), and (5) the conviction that human evolution has been a moral ascent from brutish evil to human righteousness.

PLATONIC TYPOLOGY

The Platonic type has already been mentioined in chapter 1 as a source of confusion for animal taxonomists and has been contrasted with the concept of the modal phenotype. Human as well as animal taxonomy has been sadly muddled by attempts to make human variation fit into a series of ideal types.

Blumenbach, whose 1795 classification of man into five "varieties"— Caucasian, Mongolian, Ethiopian, American, and Malay—has dominated popular thinking for nearly two centuries, was careful to point out that the "innumerable varieties of mankind run into one another by insensible degrees." His discussion of the subject of human races is surprisingly similar to modern biological concepts. He gave careful consideration to the argument that some populations of men constituted separate species, and he could find no evidence to justify such a contention. He recognized that his own classification was "arbitrary." He described his five varieties on the basis of hair, skin and eye color, and the form of the features of living individuals. His descriptions dealt almost entirely with the head (Blumenbach 1795, pp. 264−69). But in addition, in the museum-drawer tradition, he designated five skulls in his collection as exemplifying the characteristics of the five human groups. The one typifying the European group was that of "a Georgian woman" (1795, p. 155), hence the name Caucasian.

Later students paid scant attention to Blumenbach's observation of the blending of human races into each other and came to put more emphasis on morphology than on populations. They tried to define human groups with greater precision; and since, as Blumenbach had pointed out, there was great variability within every group, they defined various subtypes into which they tried to fit the observed variability. For the European populations one began to hear of Nordic, Alpine, Mediterranean, Baltic, Dinaric, and God alone knows how many other races and subraces. These were pictured as ideal morphological types that rarely existed in pure form—very much like Lorenz's butterfly. Since individuals approximating different types and combinations of types kept turning up in the same population and even among the offspring of a single couple, this apotheosis of ideal morphology all but divorced race from biology. It was as if there were a set of disembodied types wandering through space and time and being repeatedly reincarnated in individuals chosen here and there at random.

111

The climax of the racial-type game played in this manner can be found in the work of E. A. Hooton (1955), who collected vast quantities of measurements and observations on the population of Ireland in the 1930s, transferred the data to punch cards, and subjected the cards to a complicated processing designed to determine the extent to which each type contributed to the population. The results, finally published in two volumes in 1955 by the Peabody Museum of Harvard University, apportioned the population into eight categories, giving the percentage of each: Pure Nordic 0.6 percent, Nordic Mediterranean 28.9 percent, East Baltic 1.1 percent, and so on. This very extensive investigation offers no meaningful conclusion. It is only a mild exaggeration to say that it evokes little reaction beyond "So what?"

Racial typology has been taken very much less seriously in scientific circles in the last three decades, but it would be reckless to prophesy that it will not at some time be revived. Undoubtedly the excesses to which racial doctrines were carried by the Nazis contributed to the disfavor with which such ideas are now regarded. That this kind of make-believe still stalks the earth, however, and has vicious consequences is shown by the following Reuters dispatch which appeared in the *New York Times* of May 3, 1967.

> Pretoria, South Africa—The Supreme Court here today upheld the classification of 11-year-old Sandra Laing as colored (mixed race) although her parents and their other children are all classified as whites.
>
> Sandra was reclassified colored by the Race Classification Board in February last year. She was said by South African officals to be a genetic throwback, showing strong nonwhite characteristics.
>
> Dismissing an application by her father, Abraham Laing, for Sandra to be classified as white, Justice Oscar Galgut said the girl might still become white under legislation now before parliament that would make descent, rather than appearance and general acceptance, the standard for race classification.

Criticism of the idea that there are clearly recognizable racial types does not imply that all human populations are uniform or that populations cannot be grouped into larger units that, among themselves, have certain similarities. What it does imply is that abstract types divorced from populations do not exist and that similar phenotypes in the same or in different populations are not, because of that similarity, closer to each other genetically than they are to other phenotypes.

THE MYTH OF PURE RACES

Since one finds in any population not one single type but several plus a large majority of intergrades, an attractive hypothesis for explaining this untidy condition is that once upon a time every type was represented by a pure race but that interbreeding has resulted in mutual "contamination." It is fatiguing even to contemplate the amount of energy expended and the number of printed pages produced in efforts either to establish this hypothesis or to explain present observations on the basis of it.

At the beginning of the nineteenth century, European students of philology discovered systematic similarities between the ancient languages of India and Iran—Sanskrit and Zend—and the major language groups of Europe—Greek, Latin, and the Romance languages developed from it, Teutonic, Slavic, and Celtic. Evidence from vocabulary, grammar, syntax, and phonetics was sufficient to convince the linguists that all these languages must have developed by gradual divergence in isolated populations from one common stem language—a process conceptually similar to biological speciation but effected by entirely different mechanisms. Since the languages developed from this common source covered most of the huge area from the Ganges valley to the lands bordering on the Bay of Biscay and the North Sea, the hypothetical root language was named Indo-European and the people who at some remote time must have spoken it were called Indo-Europoeans.

From 1800 to the present day, all the results of linguistic research have strengthened the hypothesis of the common origin of the Indo-European languages. In fact, some languages formerly thought to be unrelated have since been shown to have the same ultimate source. But the linguistic data were also used as the basis for another hypothesis for which there is little, if any, substantial evidence. This was that the people who spoke Indo-European constituted a single "pure" race. This hypothetical race came to be known as Aryan, from the name that the Sanskrit-speaking invaders of India called themselves. Throughout the nineteenth century this fanciful theory was elaborated and embellished and used to support a variety of claims to national grandeur and superiority. It was used not only by the Nazis but at one time or another by English, Greek, French, and pre-Nazi German apologists as well. The Aryan story has been told in many different versions, but the general picture is that the vigorous and superior speakers of Indo-European went out as conquerors from their homeland, bringing with them not only their language but also their culture, intelligence, ethical sense, and superior genes. These noble people, according to the myth, were contaminated to a greater or lesser extent by intermarriage with the people among whom they settled, but exact descriptions of the uncontaminated Aryans and the means of recognizing the exact nature and degree of contamination are subjects on which there is considerable difference of opinion.

There is, of course, no evidence that the speakers of Indo-European were any more uniform genetically than any other population. We know that all populations contain variability, and there is no reason to think that the hypothetical Aryans were any less variable than most peoples. Furthermore, we know that people of very different origin can speak the same language. Hence, there is no reason to assume that two populations speaking similar languages are necessarily similar genetically.

A glance at some of the linguistic changes that have taken place in historic time yields convincing evidence that it is an error to identify language and alleles. In Anatolia in Hellenic times Greek was spoken only along the Ionian coast. In the Hellenistic period Greek spread eastward into the interior and took the place of the many different local languages, so that in Roman and Medieval times the language of Anatolia was Greek. With the Turkish conquest in the late Middle Ages Greek was replaced by Turkish, although Greek held on as the spoken language of the Ionian cities until the systematic population exchanges of the 1920s. The changes in the speech of the populations of the interior from pre-Greek languages to Greek in the Hellenistic period and from Greek to Turkish in the late Medieval period took place without any corresponding change in the genetic character of the population.

In Roman times the populations of Gaul and Spain gave up their Celtic and other languages in favor of Latin. The populations were latinized; the territory was not colonized by Romans. But in northeastern Spain and southwestern France a pre-Latin language, Basque, has held on to the present day. Today the people of Asturias to the west of the Basque country speak a dialect of Spanish, a language of the Indo-European family derived from Latin. Before the Roman period the Asturians undoubtedly spoke a pre-Latin language. But there is no reason to think that the linguistic change from pre-Latin to Latin to Spanish has in any way changed the degree of genetic similarity between Asturians and Basques.

The myth of the Aryan has not, of course, been the only case of postulated race purity, but it has probably been the chief reason for the persistence of the idea of pure races in modern Western writing. In its most naive form the doctrine is now pretty generally rejected by serious students. But bits and pieces of the idea still keep turning up even in serious scientific writing. In the Rh blood group system there is an allele generally called Rh-negative which is fairly common in Europe but rare in the rest of the world. When, in the 1940s, it was found that this allele had a frequency of more than 50 percent among the Basques, the suggestion was immediately made that the Basques were the remnants of an earlier race that had had a very high frequency of Rh-negative alleles and that the presence of this allele in other European populations might be the result of interbreeding between Indo-Europeans and the earlier population of which the Basques—who still speak a non-Indo-European language—are the remnants. It is, of course, impossible to disprove that the frequency of the Rh-negative allele in northeastern Spain was once higher than it now is—even approaching 100 percent—and that the spread of the allele from this earlier population to the later immigrants into Europe explains its intermediate frequency in western and central Europe. But one can just as well argue that some local influence has given a selective advantage to the allele and that the selective process has gone farther in the Basque country than elsewhere. Or, one might postulate that the frequency was formerly higher throughout Europe and that it has fallen more rapidly outside the Basque area. Although the allele is rare outside Europe, it exists in China, India, Japan, and among the Bantu of South Africa. Since it is present in such widely separated populations, it could build up in frequency anywhere that it happened to have a selective advantage; or, on the founders' principle, it might have a high frequency in a population that had expanded rapidly from a small group in which the allele happened to be carried by a large proportion of the individuals. There is less reason to think that Rh-negative alleles have diffused from an early European population to such remote points as Japan and South Africa than there is to assume that the allele occurs throughout the whole species and may build up to high frequency whenever local conditions favor it.

Since the age of discovery was set off in the fifteenth century by the Portuguese mariners setting out in their galleons for the remote corners of the globe, human mobility has increased steadily to the present jet age in which few points on the earth are more than forty-eight hours from each other. This change has had the effect of bringing the most diverse and remote people into face-to-face contact with ever greater frequency. But before the modern age began human communities did not exist in isolation. There were always neighbors, over the hills or across the water, and contacts must have been continuous even when they were not frequent. In the absence of profoundly effective barriers populations showed clines of physical differences and widespread diffusion of techniques; both argue for a constant traffic in ideas and in genes from earliest times though at some times and in some places it may have been sluggish. This interchange was always sufficient to keep the species one and this was enough to prevent the development of "pure races."

Recent studies of the now extinct native population of Tasmania make a good case for a cultural retrogression there as a result of complete isolation (Diamond 1978). Tasmania was cut off from Australia by the rising waters in Bass Strait at the end of the Pleistocene, some 12,000 years ago. When Europeans came to the island late in the eighteenth century, they found a population living in very primitive stone-age conditions. European competition proved to be too much for them and by 1876 the last one had died. Excavations show that between their original isolation and the arrival of the Europeans, Tasmanian culture declined. The techniques of fishing and of making bone tools, which had flourished earlier, died out and life became more impoverished. The population appears to have been too small to maintain its culture in the absence of stimulating contact with other people. Populations had existed on smaller neighboring islands and had died out completely before the eighteenth century. It looks as though the Tasmanians may have been headed for extinction even if the Europeans had never discovered them. Apparently any population small enough and with few enough outside contacts for a long enough time to have approximated a "pure race" would have been unlikely to survive. The notion that the remote past was tidier and more attractive than the present is an ancient one, akin to the assumption of a golden age. It is obviously attractive to many people but there is little evidence for it.

RACE AND CLASS

Man as a biological species is polymorphic; individuals vary in phenotype. As a cultural animal man is polystatic; society assigns to every individual a status that, to a greater or lesser extent, determines his occupation, his mode of life, and his share of the world's goods. Physical polymorphism has a genetic component. Physical traits tend to run in families, although with considerable individual variation. Since the status of a family is generally determined by that of the parent who heads it—usually the father in Western cultures, children raised in the family are associated with his status and as adults usually wind up with a status very similar to his. Status, therefore, also tends to run in families. It is easy to see how one can come to think of status as hereditary in the same sense as phenotypic similarity although it is actually passed on as a social inheritance.

Human beings are seldom indifferent to each other. Their interactions range all the way from destructive attack to passionate attachment, and not infrequently the two extremes are intermingled. A widespread human reaction is an uneasiness about differences in status, generating arguments to prove that such inequalities flow from the nature of things and hence are not injustices that can be blamed on the selfishness of the more fortunate. Historically, this is one of the oldest of ethical problems.

Plato found this problem so difficult that he admitted hesitancy in discussing it; his own solution embarrassed him. In his ideal state all three classes—Guardians, Auxiliaries, and Common People—were to be persuaded to believe a myth: that their creator had fashioned them of three different substances—gold, silver, and baser metals, respectively—and that failure of any individual to adhere to his station would be contrary to the nature of things. But in what looks like prescience of polygenic inheritance, Plato warned that

> though children will commonly resemble their parents, occasionally a
> silver child will be born of golden parents, or a golden child of silver

parents, and so on. Therefore the first and most important of God's commandments to the Rulers is that they must exercise their function as Guardians with particular care in watching the mixture of metals in the characters of the children. If one of their own children has bronze or iron in its makeup, they must harden their hearts, and degrade it to the ranks of the industrial and agricultural class where it properly belongs: similarly, if a child of this class is born with gold or silver in its nature, they will promote it appropriately to be a Guardian or an Auxiliary (Plato 1955, p. 160).

The more pragmatic Aristotle attempted to solve the same problem without resort to poetry. From observation he concluded that "by nature some are free, others slaves" (1962, p. 34). He found it difficult to justify the enslavement of Greek by Greek as a result of conquest, but he found the enslavement of barbarian by Greek much less troublesome.

Throughout human history, ambivalence on the question of the hereditary nature of class differences has been the rule. Time after time a society has been set up by an invading group that became a ruling aristocracy—the Achaeans in Mycaenean Greece, the Aryans in India, the Franks in Roman Gaul, the Normans in England, to name a few famous instances. This phenomenon generally has given color to the theory that social classes are hereditarily different. In fact, these conquering invaders were often not very different genetically from the people they conquered, and in general they did not follow a very rigorous program for retaining their genetic distinctness. There is no doubt that in Europe the aristocracy and the peasantry have not constituted a single, randomly breeding population, but the genetic differences between them are probably no greater than the differences between populations of different localities, such as Tuscany and Calabria in Italy or Cornwall and Yorkshire in England.

As Ashley Montagu (1964, p. 125) has astutely pointed out, in the countries of western Europe where such phenotypic differences as exist are mild and continuous as compared with the differences between Spaniard and Indian in Mexico, Boer and Bantu in South Africa, or African and European in the United States, social classes are regarded as races, genetically distinct populations. Nineteenth- and twentieth-century literature is full of unselfconscious allusions to the hereditary differences between classes. Montagu cites references to the races of peasants and landlords of southern Italy in Ignazio Silone's novel *Bread and Wine*. Benjamin Disraeli published a novel dealing with radical political agitation in England in the 1840s. He called it *Sybil, or the Two Nations*, and the two nations were the rich and the poor.

In a biography of Lord Curzon, Harold Nicolson related a story that beautifully illustrates the degree to which race and class are commonly identified. Curzon was one of the most typically aristocratic of British statesmen. He served as viceroy of India in the 1890s and as foreign minister of Britain in the period following World War I.

> Behind the lines in Flanders [Nicolson wrote] was a large brewery in the vats of which the private soldiers would bathe on returning from the trenches. Curzon was taken to see this dantesque exhibit. He watched with interest those hundred naked figures disporting themselves in the steam. "Dear me!" he said, "I had no conception that the lower classes had such white skins." Curzon would deny the authenticity of this story, but he loved it nonetheless (Nicolson 1934, p. 48n).

This story is particularly revealing for the unquestioned assumptions it uncovers: first, the class stratification of the British Army (of course, the private soldiers were from the lower classes); second, the lower classes were expected to be swarthier than aristocrats; and finally, the denial and simultaneous approval of the story (the hereditary differences between classes constitute a delicate subject that a gentleman hesitates to bring up, but, of course, the facts are undeniable). One senses in this diplomatic repudiation something close to Plato's embarrassment over his founders' myth.

Of course, belief in the importance of biological descent is by no means confined to the aristocracy. The notion of "blood" as a powerful influence in determining individual status and worth is widespread and ancient in all groups and classes. It is the basis of kinship systems and usually of the law of inheritance and of citizenship (*jus sanguinis*). But racial inequality of ability as a political doctrine arose, in fact, not out of conflict between the very dissimilar peoples of the earth, but from the class differences of western Europe. The racial superiority of the Aryans was expounded as a doctrine by Count de Gobineau, a French aristocrat, in *Essai sur l'inégalité des races humaines*, published in Paris in the 1850s. He argued that the principle of equality, the cornerstone of the democratic movement of the nineteenth century, was a political evil because the aristocrats, descendants of superior progenitors, were capable of ruling, whereas the lower classes were of inferior origin and were therefore politically incompetent.

De Gobineau's work was published in translation in the United States before the Civil War and was used by Southern apologists for slavery as ammunition against the abolitionists. Obviously, the superiority of the Aryan aristocrats to the peasantry and the proletariat of Europe only clarified the greater gap between European and the less-well-endowed non-Europeans. In the period following the American Civil War the wretched position of the American Negro and the triumph of colonialism in Asia and Africa pushed the question of equality between Europeans and non-Europeans so far from the realities of politics that it was not a very lively topic of discussion in the West.

The followers of de Gobineau in Europe and America came to emphasize the superiority of the "Nordics," a term that like Aryan proved flexible, the exact definition depending on the user. In the mid-nineteenth century in the United States, people of English descent proclaimed their superiority over the Celts. So distinguished a New Englander as Charles Francis Adams, Jr., wrote (1892, p. 957), "Quick of impulse, sympathetic, ignorant and credulous, the Irish race have as few elements in common with the native New Englanders as one race of men well can have with another." By the 1920s one heard less of the racial inferiority of the Irish and more of the inferior germ plasm from eastern Europe which had filled the slums of American cities. The American Immigration Act of 1924 which introduced the "National Origins" policy was deliberately designed to put a stop to what was regarded as the racial degradation of the American population. In Europe the idea of the racial superiority of the aristocratic classes persisted.

The all-pervasiveness of the identity of socioeconomic and racial differences is beautifully illustrated by the following quotation:

> The bus ride from Rotherhite to more familiar parts of London (Kensington, Hyde Park, Oxford Street) takes well over an hour. It winds past miles and miles of workers' flats with such incongruous-sounding names as Devon Mansions, Cornwall Homes. Five or six stories high, these gloomy red-brick structures house a shorter and paler race of people than the inhabitants of London's West End. In appearance, dress, and speech they form so radical a contrast as to give the impression of a different ethnic group (Mitford 1960, p. 172).

This was written by Jessica Mitford. No racist herself, she deplored the racism of her sister Unity, who for a time was in Hitler's inner circle; and she regarded her father, Lord Redesdale, as socially irresponsible because "When one of our cousins married an Argentinian of pure Spanish descent, he commented, 'I hear that Robin's married a black.' " Yet, in describing the working people of London, she wrote "shorter and paler race."

The diagnostic characters of social classes vary, but while physical phenotype is important where it exists, it is by no means necessary. Behavioral characters are much more compelling. Probably most significant is language. Pronunciation, vocabulary, inflection, and locution are telltale evidence not only of present status but also of personal history. We could not have had *Pygmalion* nor its successor, *My Fair Lady*, without this social phenomenon. After language comes dress—the type of collar, blue or white; the hat or cap; the shoes; the countless flamboyancies or subtleties of style depending on the time and place—all contribute to the determination of the social pigeonhole. Then, also, there is behavior, posture, gait, forms of polite response, table manners. Finally there are attitudes and beliefs. All these characters are important for mutual evaluation whenever there is an encounter between two strangers.

To exactly what extent these different characters can be traced to hereditary components is not easy to say. All of them—even physical phenotype to a considerable degree—can be counterfeited by a competent actor and a makeup artist. But, in fact, we know that most of these characters are strongly influenced by culture. This is often true even of physical phenotype; for example, whether at a given age individuals are under- or overweight. Society prescribes these distinguishing marks and inculcates them by constant pressure and drill in order to facilitate keeping track of status, the imperfect classification developed to carry out the objective of Plato's founding myth—keeping people in their place.

The difficulty of making a clear distinction between class and race is still with us. Those who emphasize racial differences in I.Q. also stress differences in I.Q. between social classes. Jensen (1969a, pp. 74−75) points out that the I.Q. scores of schoolchildren are correlated with the socioeconomic status of their parents. Doubting that socioeconomic status as such can determine the value of the I.Q., he concludes that there must be differences in genetic endowment for intelligence between socioeconomic groups. At the same time he considers social class as distinct from race. But unless social classes are reconstituted anew in every generation by screening and classifying all individuals according to I.Q.—which, of course, they are not—it is hard to draw a sharp line between race and class. Classes are populations between which interbreeding is less likely than it is within each class; and to the extent that the two classes differ genetically, they do constitute ecological races as was pointed out in chapter 1.

But a hopeless confusion arises if we use as a measure of genetic difference the culturally inculcated elements of behavior which are badges of class membership. The only evidence we have of the heritability of the I.Q. comes from comparisons between relatives. Since we know so little of the details of environmental influences on the development of the I.Q. we are completely unable to randomize environmental factors in such studies, and without such randomization the heritability figures obtained must be regarded with great skepticism. The ghost of Plato's myth keeps coming back. Are the differences in I.Q. between social classes and the heritability figures claimed real measures of genetic difference, or do they arise from the high efficiency of culture in stamping the individual with marks of his status? Many of the other marks of status are extremely subtle and we know little of exactly how they are acquired.

The confusion here certainly is in part because of a collision between an aristocratic and a balanced view of population structure. Those who emphasize the hereditary component of the I.Q. deplore the fact that of our present population only 2¼ percent have I.Q.s of 130 or above and that the population mean is 30 points lower. They yearn for their ideal of a genetically improved population, where the mean will be the present 130 and the standard deviation will have been drastically reduced. In this new Utopia there would be in terms of present standards no dull people and few "normal" ones. But the dreaming of such a dream betrays a lack of understanding of polygenic systems and genetic homeostasis. The mean I.Q. in human populations is where it is not because natural selection has failed to peg it at 130, but because in relation to all other characters the present mean is the one consistent with maximum adaptiveness of the population. It could probably be pushed up by artifical selection, but this would almost certainly produce maladaptive concomitants. On termination of the selective program, the mean would probably decline toward its former value and the maladaptive symptoms would fade out.

We have to say probably because, fortunately, no artificial selection experiment for raising the I.Q. has been carried out on a human population. But some very interesting selection experiments on behavioral characters in *Drosophila* were made by Dobzhansky and Spassky (1967). Using an apparatus composed of a set of glass tubes that fork four times, they were able to divide a group of flies into sixteen classes depending on how many right- or left-hand turns they had made during their progess through the tubes. If at each fork one tube is covered with opaque material, the flies are given choices between light and dark and the classification becomes one based on positive or negative phototaxis (movement toward light). If the apparatus is turned on its side and all tubes are covered, the classification is a measure of response to gravity (geotaxis).

Laboratory populations turn out to be essentially neutral with respect to both types of behavior, but if one breeds from flies that wind up at the ends of the distribution, one can raise or lower the mean performance for either character. The heritability is not high—8 to 10 percent for phototaxis and only about 3 percent for geotaxis—but continued selection produced response for some seventeen generations. At this point selection was terminated, and each population was divided into two, one of which was allowed to breed without further selection and the other subjected to reverse selection. In every case the program of reverse selection brought the mean back to its original position in about half the number of generations it had taken to drive it away. In populations where no selection was practiced, the mean also drifted back, but at a slower rate.

These results are consistent with what has been found in other artificial selection experiments, and such data form the basis on which our theory of polygenic inheritance and genetic homeostasis is constructed. Dobzhansky, Spassky, and Sved (1969) carried the experiment somewhat further with very interesting results. From a population that was being selected for positive phototaxis, some flies from the bottom of the distribution—rejects that were not to be used as parents of the next generation—were introduced into another population that was not being selected at all. A positive response occurred not only in the selected population but also—although at a much slower rate—in the unselected population that received the rejects as immigrants.

These results may appear paradoxical, but they are quite consistent with polygenic inheritance. The mean is where it is because it is associated with high adaptedness; and alleles that, in the proper combinations, can give high values are present in individuals in the lower part of the distribution. Failure to understand this type of inheritance leads to an overemphasis on the importance of hereditary components in behavioral characters and to a distorted picture of human variation which exaggerates the hereditary differences between both classes and races.

EMOTIONAL RESPONSES TO RACIAL DIFFERENCES

In December 1918 the *National Geographic Magazine* published an article entitled "The Races of Europe" by Edwin A. Grosvenor. The "races" listed and described are primarily linguistic groups, as becomes clear from a glance at the colored foldout map that accompanies the article. Scattered through the seventy pages of text are numerous illustrations of individuals of different races. There can be little doubt that these illustrations had greater impact on the reader than did the text, and it is extremely interesting to try to imagine the kind of picture of the people of Europe the average American reader of the time would have got if these illustrations were his sole source of information.

One photograph of a Hungarian swineherd shows a lonely figure in a wide-brimmed, low-crowned hat; a long, heavy cape; and high boots; he contemplates a wide expanse of the Danube with the opposite shore low in the distance. In the foreground are nineteen pigs. Another illustration is a black-and-white reproduction of Raphael's Sistine Madonna to illustrate the "Greco-Latin face and figure." Another photograph with the caption "A Gateload of Smiles in Sweden" shows twelve children, both boys and girls, ranging in age from perhaps six to eleven, dressed in neat but unchic rustic garb, sitting on a gate with a rural background showing a fence, a small building, and birch trees. The scene is somewhat reminiscent of John Greenleaf Whittier. Another photograph of a superannuated, wrinkled man in a cutaway coat, knee breeches, and a stovepipe hat is captioned "An Irishman of the Old School." In the doorway of the cottage before which he is standing is an equally elderly woman in a bonnet and shawl. The final picture in the article is another reproduction of a painting. It shows a young woman in classical draperies leaning against a low marble wall with a bouquet of flowers held carelessly in one hand. The caption is

"Daughter of a Race of Empire Builders." The empire referred to was, of course, the British.

Grosvenor's article was timely, because at Paris that winter the boundaries of post-World War I Europe were being drawn and, in accordance with the Wilsonian principle of self-determination, previously submerged linguistic groups—the Poles, the Czechs, and the Yugoslavs—were being set up as independent states. But the illustrations threw no light whatever on the similarities or differences between the various populations of Europe and were positively misleading in that they made no suggestion of the cultural similarities between early twentieth-century Budapest, Rome, Stockholm, Dublin, and London.

If one were attempting to describe the different subspecies of the American gray squirrel, one would become familiar, through skins in museums, notes, drawings and photographs by other investigators, and one's own field work, with the amount of variation within and between populations and then would describe and illustrate it as vividly as possible. One would not show a hairless, newborn form from one area, a lactating female from another, and a moribund specimen from a third. One would try to show the various age and sex groups in each population and something of how much variability there was within each. But however one illustrated one's work on the gray squirrel, one would know that the reader would look at it as an outsider. He might find the squirrels beautiful, ugly, attractive, or cute; but he would not react to them as potential companions, rivals, victims, or bedfellows.

When one human being contemplates another, in the flesh or in representation, his reactions are different from those he has with respect to any other animal or object. There is a unique type of emotional involvement. Blumenbach, in describing the "Caucasian variety" of mankind, wrote (1795, p. 269), "I have taken the name of this variety from Mount Caucasus . . . because its neighborhood . . . produces the most beautiful race of men, I mean the Georgian." Then, in a footnote, he quoted from another author: "The blood of Georgia is the best of the East, and perhaps in the world. I have not observed a single ugly face in that country, in either sex; but I have seen angelical ones. Nature has there lavished upon the women beauties which are not to be seen elsewhere. I consider it to be impossible to look at them without loving them." Blumenbach did not illustrate his book with pictures of living people. He figured only drawings of skulls. The "Caucasian variety" was represented by a skull identified in the legend as "a Georgian woman." In the text he refers to this figure as "the most beautiful form of the skull."

One simply cannot illustrate a work on human races without using loaded dice. Ugly, old models evoke a negative reaction; young, attractive ones create a feeling of identification and approval. But the effect of physical phenotype is strongly influenced by cultural modifications. Dress, hairdo, beard, tattooing, bone through the nose—all such cultural additions to the phenotype have important influences on the emotional reaction of the viewer. In dress, even the modishness has a powerful influence. What looked very smart ten years ago looks ludicrous today, and it is next to impossible to suppress the overtones of such reactions. One would think that this problem might be avoided by presenting the representatives of different races unclothed, but while this procedure might get rid of the distractions caused by dress, it would bring in others that would be no less subjective. Carlton S. Coon's *Living Races of Man* (1965) has a set of plates showing individuals of different races. If one looks at plate 5, "A Yukaghir," an old wrinkled man with scraggly beard and long stringy hair, and then at plate 76, "A Hawaiian Woman of Hawaiian, English, and Chinese Ancestry," an airline hostess in what still appears to be a smart uniform and skillful makeup, one is immediately aware that the information conveyed by the two photographs is in no way relevant to the differences between the two populations.

But not only do people react to each other emotionally and not only is the emotion sometimes erotic, but differences in physical phenotype—real or fancied—often arouse deep-seated anxiety and guilt and produce bizarrely irrational behavior. One of the oldest and most fundamental problems of human culture has been the socialization of the sex drive and of sexual activity. Different cultures have used many different ways of attempting to solve it, but one common element in all solutions has been a tendency to follow a middle course in mating, between incest on the one hand and extreme exogamy on the other. The near universality of these practices is probably not based primarily on biological considerations. It is much more likely that they have developed as effective means of insuring group cohesion and reducing intragroup conflict. But in any case they are ancient and are usually supported by intense emotion. Consequently, not merely the practice, but even the fantasy of either incest or sexual activity with a partner of very different origin commonly arouse feelings of guilt.

Human culture generally appears to condition its members to feel that they can minimize their guilt concerning sex if they avoid incest but at the same time keep within their race. It is interesting that in English the word *breed* means both "race" and "to procreate": "to breed" is permissible within "the breed." In German the word *Geschlecht* means both "sex" and "race." And in Portuguese the word *casta*, which means "race," derives directly from the Latin adjective meaning "morally pure." Whatever the terms, the concepts of race and sex have a way of becoming inextricably entangled. On the one side, the members of a different race stand as symbols of temptation to sinful debauchery; on the other, they represent demonic schemers plotting rape or seduction of one's virtuous confreres. Most of the noxious distortions invoked to justify doctrines of racial superiority and practices of racial discrimination and persecution have been rationalizations based on these tangled, hidden motives. In such a frame of mind no one is capable of making an objective evaluation.

The results of these dark forces are to be found in their ugliest manifestation in a speech made in the United States Senate by Senator Ben Tillman of South Carolina on January 21, 1907, (Cong. Rec. 59th Cong. 2nd Sess. pp. 1440−44) which has been used by Martin Duberman (1964, pp. 62−63) with such great effectiveness in his *In White America*.

> engulfed, as it were, in a black flood of semi Barbarians [the Senator said], . . . the white women of the South are in a state of siege . . . [he then evoked the picture of] a fair young daughter just budding into womanhood. . . Some lurking demon who has watched for the opportunity seizes her; she is choked or beaten into insensibility and ravished. . . . Is it any wonder that the whole countryside rises as one man and with set, stern faces seek the brute who has wrought this infamy? . . . It is idle to reason about it; it is idle to preach about it. Our brains reel under the staggering blow and hot blood surges to the heart. Civilization peels off us, any and all of us who are men, and we revert to the original savage type whose impulses under any and all such circumstances has always been to "kill, kill, kill."

BIOLOGY AND MORAL VALUES

Blumenbach believed that the "Caucasian variety" of mankind was the original form and that it had diverged

> into two, most remote and very different from each other; on the one side . . . the Ethiopian and on the other into the Mongolian. The remaining two occupy the intermediate positions between that primeval one and these two extreme varieties; that is the American between the Caucasian and the Mongolian; the Malay between the Caucasian and the Ethiopian (1795, pp. 264−65).

The four non-Caucasian varieties, he thought, had arisen by "degeneration" from the original white, "since it is very easy for that to degenerate into brown, but very much more difficult for dark to become white" (p. 269). We have to remember that Blumenbach never suspected that one species could change into another. By "degeneration" he meant merely a hereditary shift in general appearance such as had clearly taken place in many domestic animals to produce different breeds. The term as he used it did not have a pejorative connotation and, of course, it did not imply what was for him an impossibility, a change of one species into another.

As Darwin's theory that natural selection could cause one species to evolve into another became more and more generally accepted in the second half of the nineteenth century, the relative positions held by human races in Blumenbach's scheme were radically altered, in fact reversed. If man had evolved from a lower animal, the ancestral form must have been apelike. There were no white apes. The white race was the most civilized. Therefore the dark races, which were culturally more primitive, were closer to the nonhuman progenitor, and the white populations of western Europe represented the latest and highest form of evolutionary progress.

This seemingly plausible description of the state of human affairs became almost unquestioned both in scientific circles and in the popular mind, and several influences caused it to be elaborated and rationalized and to give the color of scientific justification to the doctrine of racial inequality. One of these influences was the ancient idea of the scale of nature. Since classical times, various authors had discussed in greater or less detail the arrangement of living things in an ascending series from the very simple to the most complex and that the individual steps in this progession were often very small. Combined with the doctrines of special creation and fixity of species, the scale of nature did not suggest that similarity of structure had resulted from common descent. It was, instead, given a mystical interpretation; the whole of the scale of nature represented the inscrutable working of the mind of the Creator.

Monkeys had long been recognized as close to man in the scale of nature. The great apes were even closer. The careful eighteenth-century systematists such as Linnaeus and Blumenbach recognized that man was a single species distinct from any of the apes, but in more popular writing and in folklore the idea was widespread that there were numerous intermediate forms between men and apes and that it was not exactly clear just where to draw the lines. The name orangutan comes from a Malay expression meaning "man of the forest." The idea that intermediate creatures closer to man than any of the known apes may exist still survives; many people have not yet written off the Abominable Snowman of the Himalayas as entirely imaginary. As Europeans learned more of the technologically primitive peoples of America, Africa, Asia, and Oceania, it was quite natural that these imperfectly known and less understood "savages" should be regarded as intermediate creatures in the scale of nature between man and the equally poorly known great apes.

With this background, it is easy to imagine the confusion that acceptance of Darwin's hypothesis of evolving species brought to the problem of classifying man. In *The Origin of Species* Darwin made only one oblique reference to the origin of man. But as soon as one gave up the dogma of special creation and belief in the fixity of species, the discrete steps in the scale of nature began to break down, and the whole structure changed from a static reflection of the ideas of the Creator to a dynamic, fluid process. It ceased to be a curio cabinet and became a living, teeming world. The question of man's place in this scheme could not be ignored, and it at once became central to the passionate debate over evolution. In a later volume, *The Descent of Man*, Darwin wrestled with the problem at length and marshaled the evidence for man's origin from an apelike creature.

But the acceptance of natural selection as the agent for restructuring and creating species still left unsolved a great many problems connected with the understanding of human variation. Nothing was known at that time of the mechanisms of heredity. There was very little understanding of the nature of human culture. The application of the new understanding of biological evolution to the human species was made by western Europeans from their own provincial point of view. Since they had achieved a higher technical and scientific level than any other people, they assumed that they represented the pinnacle of evolutionary development. They did not distinguish clearly between the biology and the culture of a population. Consequently, they thought in terms of a single line of development for human populations. Those peoples who had not yet reached the highest point were at a position farther back on the single highway and hence were lower in the scale.

The smugness and arrogance with which many of the most reputable nineteenth-century scientists of western Europe assumed their own superiority is, even to our own still benighted minds, close to incredible. For example, in 1869 Francis Galton, one of the most brilliant of Victorian Englishmen, wrote in *Hereditary Genius*:

> A native chief has as good an education in the art of ruling men, as can be desired; he is continually exercised in personal government, and usually maintains his place by the ascendancy of his character, shown every day over his subjects and rivals. A traveller in wild countries also fills, to a certain degree, the position of a commander, and has to confront native chiefs at every inhabited place. The result is familiar enough—the white traveller almost invariably holds his own in their presence. It is seldom that we hear of a white traveller meeting with a black chief whom he feels to be the better man (Galton 1962, p. 394).

And T. H. Huxley, the great protagonist of Darwin's theory, wrote in 1865:

> It may be quite true that some negroes are better than some white men; but no rational man, cognizant of the facts, believes that the average negro is the equal, still less the superior, of the average white man. And, if this be

true, it is simply incredible that, when all his disabilities are removed, and our prognathous relative has a fair field and no favor, as well as no oppressor, he will be able to compete successfully with his bigger brained and smaller jawed rival, in a contest which is to be carried on by thoughts and not by bites. The highest places in the hierarchy of civilization will assuredly not be within the reach of our dusky cousins, though it is by no means necessary that they should be restricted to the lowest (Huxley 1910).

Nineteenth- and early twentieth-century Westerners regarded such out-of-the-way peoples as Bushmen, Hottentots, and Tasmanians as being—both biologically and culturally—at an evolutionary stage much closer to that of man's animal ancestors than to that of the Caucasians of western Europe who had passed through that stage many thousands of years earlier. All other peoples, it was assumed, could be fitted into intermediate stages along a linear continuum between the two extremes.

In recent years the question of the relation of the human to the ape has become complicated by developments in molecular biology. Techniques have been devised to measure "genetic distance" between different species (King and Wilson 1975). This procedure is based on the number of amino acid substitutions in similar proteins or on the number of differences in nucleotide sequences in the DNA. The protein differences have been measured in three ways. Where proteins have been completely sequenced, they can be compared directly and the number of positions where amino acids differ can be counted. But sequencing is time-consuming and expensive and there are two other methods that can be used for unsequenced proteins.

The first of these is immunological. One protein can be purified and an antibody against it prepared by injecting it into a rabbit or some other mammal. This antibody, obtained by bleeding the animal and extracting the serum, can be used to measure its difference in affinity for the protein against which it was formed and the similar protein of another species. The second method employs electrophoresis. Protein can be put on a gel, subjected to an electric field, and the distance it moves within a given time measured. Pairs of proteins can be compared in this way.

Differences in nucleotide sequences can be measured by taking DNA from two species, dissociating the two strands by heating, mixing the two resulting solutions of single strands and allowing them to anneal. If there are substantial differences in sequence, the annealed double strands of unlike origin will dissociate at a lower temperature than that required to cause the original like strands to come apart.

Studies of chromosome morphology show surprisingly small differences between man and chimpanzee karyotypes (Yunis, Sawyer, and Dunham 1980). The cytological study of chromosomes, however, is at considerable remove from the molecular level and no quantitative measure of the difference has been worked out. When the proteins and DNA of humans and chimpanzees were compared, the distance was found to be extremely small. Differences between similar species of squirrels turn out to be several times greater than the human-chimpanzee difference and differences between similar frogs are twenty to thirty times as large. In fact the difference based on proteins was so small that it produced a lively argument between the molecular evolutionists and students of the human fossil record (Maynard Smith 1975). A broad comparison of the values for genetic distance between groups of animals of widely different degrees of relationship suggested that over long periods of time amino acid changes accumulated at a constant rate. This led to the theory that from these molecular differences one could read the number of years since two animals had a common ancestor. On the basis of this theory, the small differences between human and chimpanzee would have accumulated in a period of only three million years. But the fossil record indicated that the two lines must have separated from fifteen to twenty million years ago.

In this argument the paleontologists have come out the winners. The fossil record, which during the past ten years has increased enormously in both information and complexity, clearly shows the human lineage as distinct from that of the apes, existing in the fossil genus *Ramapithecus*, at least as far back as the middle Miocene, some fifteen million years ago (Leakey and Lewin 1978). The molecular evolutionists, on their part, have relaxed their contention that proteins and morphology necessarily change at the same rate. King and Wilson (1975) have pointed out that in spite of the small molecular differences between chimpanzee and human, there is a much wider morphological gap and a still wider behavioral one. They suggest that these changes have come about as a result of regulatory mutations. Since their paper was written, the discoveries of intervening sequences and RNA processing—already discussed in some detail in chapter 3—have revealed the enormous potentiality for changes in regulatory messages and have made it easier to understand why amino acid substitutions alone are of limited value in the measurement of genetic distance.

There is, of course, no evidence that any existing human population is any closer to the apes than any other. When the measurements of protein differences within the human species are compared with those between ape and human, the latter are from twenty-five to sixty times as great as any difference between two human populations, and neither Caucasians, Black Africans, nor Japanese are any nearer to the chimpanzee than either of the others. Morphologically, the chimpanzee is very different from the human. Here the differences are much greater than they are in many cases of pairs of species that show much greater differences based on protein measurements. The fundamental characters that distinguish the apes from the hominids, such as the ability of the great toe to function like a thumb, the large canine teeth, and the pelvis unadapted for walking erect were all lost by the hominid line many millions of years ago and they appear in no living human populations.

Superficial similarities between apes and men appear randomly in different human groups. So far as pigmentation is concerned, the human blacks are most apelike, but the chimpanzee is less black than some Africans. In general the apes have fine, straight, and profuse hair. Among humans, these characters are most closely approximated in the Caucasians. Among blacks the hair is coiled and on the body it is sparse; among Mongolians it is coarse, straight, and sparse. Apes have thin lips, another character most closely approached among the Caucasians. In man the spine is curved inward in the small of the back; in the ape it is not. In man this character is most pronounced and least apelike in African blacks. No human populations is in any real sense more apelike than any other. This applies both to physique and to culture. With respect to culture, the gap between man and ape is as clear and as deep as it is with physique. No human population can exist without culture; no population of apes is capable of acquiring human culture.

In addition to giving inferior positions on biological and cultural scales to living primitive peoples, the Westerners have also rated the primitives very low in morality. The Western white was "the better man." This moral element in the Western attitude was curiously strengthened by the guilt carried uneasily by many Darwinians because of their contribution to the destruction of Christian cosmology. To a great extent the argument over Darwinism became polarized as a contest between *Origin of Species* and *Genesis*. One had to make one's choice. The evolutionists were very sensitive to the charge that acceptance of man's animal origin would undermine all moral authority and render inevitable the triumph of sin and bestiality. There is an eloquent and touching passage in T. H. Huxley's *Lectures and Lay Sermons* which attempts to place natural selection on the empty throne of the Deity.

> Yet it is a very plain and elementary truth that the life, the fortune, and the happiness of every one of us, and, more or less, of those who are connected with us to depend upon our knowing something of the rules of a game more difficult and complicated than chess. It is a game which has been played for untold ages, every man and woman of us being one of the two players in a game of his or her own. The chess-board is the world, the pieces are the phenomena of the universe, the rules of the game what we call the laws of nature. The player on the other side is hidden from us. We know that his play is always fair, just and patient. But also we know to our own cost that he never overlooks a mistake, or makes the smallest allowance for ignorance. To the man who plays well, the highest stakes are paid with that sort of overflowing generosity with which the strong show delight in strength.

And one who plays ill is check-mated—without haste, but without re-morse (Huxley 1910, p. 58).

The father figure that Huxley describes is perhaps less horrendous than George Orwell's Big Brother. He is fair and patient. But there is little suggestion of compassion in his makeup. The stern moral tone of this quotation remained typical of many Darwinians well into the twentieth century, and it supported the belief that moral progress was a part of biological evolution. It seemed axiomatic that Western civilization was technically, culturally, and *morally* superior to less technically proficient cultures and that this difference arose from the supposition, which they did not question, that the more "primitive" peoples had not yet journeyed so far from their brutish origins. Twenty-three years after the original publication of *Hereditary Genius*, Galton wrote in the preface to the 1892 edition:

> There is nothing either in the history of domestic animals or in that of evolution to make us doubt that a race of sane men may be formed, who shall be as much superior mentally and morally to the modern European, as the modern European is to the lowest of the Negro Races (Galton 1962, p. 27).

The assumption of the brutishness and low morality of different people has been clearly shown in the attempts made by paleontologists to reconstruct fossil man. The flesh and hair on such reconstructions have to be filled in by resorting to the imagination, but the imagination has generally worked in only one direction. Almost all the sketches and plaster models that have been made of early man, from Java man to Neanderthal, have shown an ugly brutish creature. Figure 10 shows this up-from-the-brute sequence in three restorations made by J. H. McGregor in the first quarter of this century of what we now call *Homo erectus*, Neanderthal man, and Cro-Magnon man. The first looks clearly brutish and stupid; the second is somewhat less so, although to head off any pretension he is given a skid-row beard; the third has a rugged dignity recalling a Roman Senator. Cro-Magnon man, who first appeared in Europe between 35,000 and 40,000 years ago, was the first to enjoy good public relations with the restorers. Some artists' sketches of his family life resemble scenes from the biblical epics of Cecil B. DeMille. The persistence of this notion that human evolution is the story of a stupid brute being outsmarted and supplanted by a clever newcomer is dramatically portrayed in a painting by Jay H. Matternes published in the Time-Life Book *Early Man* (Howell 1968). It shows an encounter between a group of males, females, and young of a backward hominid (*Australopithecus robustus*) and two males of a more progressive species (*Australopithecus africanus*), the latter equipped with stone weapons of the latest paleolithic technology. The

FIGURE 10. Three restorations of prehistoric man by J. H. McGregor. From the left: Java ape-man (*Homo erectus*), Neanderthal man, Cro-Magnon man. A progression from brutishness to nobility is obvious. (Photographs courtesy of the American Museum of Natural History.)

backward males resort to a rear guard action, picking up whatever unimproved missiles lie at hand, while the women and children scamper off. Of course, no one can say that such an encounter never occurred, but neither is it well established that human evolution was the result of a series of prehistoric rumbles.

Since Neanderthal remains in Europe and western Asia run out about forty thousand years ago and are succeeded by those of modern *Homo sapiens* represented by the Cro-Magnon, many people have assumed that this shift was the result of a violent confrontation between the primitive, brutal Neanderthal and the more advanced, more intelligent, and more worthy Cro-Magnon. Although no hard evidence for such a clash exists (Trinkaus and Howells 1979), it has been luridly described in fiction at least four times: by Vardis Fisher in *The Golden Rooms* (1944), by William Golding in *The Inheritors* (1955), and more recently by Jean M. Auel in *The Clan of the Cave Bear* (1980) and by Björn Kurtén in *Dance of the Tiger* (1980). In all these novels Neanderthal succumbs because of inherent inferiority. Kurtén, who is a competent, recognized paleontologist, attempts a more sophisticated version of the tale. He makes his Neanderthals white and gives them a culture based on extreme ritual politeness, but they are morphologically incapable of producing the musical sounds of Cro-Magnon speech. The Neanderthals are enamored of the more attractive Cro-Magnons and prefer them as mates to their own kind. The hybrid offspring are sterile and so in a few generations the Neanderthals are no more. It seems as difficult to eradicate racism from prehistory as it is to get it out of contemporary life.

In a very recent and interesting analysis of the Neanderthal enigma Valerius Geist (1981) finds these people neither primitive nor inferior but more highly specialized in physique and anatomy than any other known human group. This specialization, he argues, was the result of selection for survival in the peculiar ecology of the periglacial environment in which human life depended on "close quarter confrontation hunting" of large, powerful herbivores. With the retreat of the glacier, the ecology changed radically and the Neanderthal specializations no longer proved adaptive. Geist's argument is based on a complex set of deductions from the fossil record, not all of which can be said to be firmly established, but it is intellectually more attractive than a simplistic dumb-brute-bested-by-smart-upstart scenario and it is evidence that this latter is by no means the only explanation of the record.

The mingling of biology with intellectual and moral worth led to the assumption that because man's biological nature made him capable of conceiving an ethic and acting ethically, it was therefore inevitable that only one ethical system—that of the West—could represent the full flowering of man's moral potentialities. This assumption is quite as invalid as was the belief of Pharaoh Psammeticos that a specific language was inherent in every child. It is still widespread, deep-seated, and hard not merely to eradicate but even to bring to the conscious level. In *Race*, an imposing, scholarly, 625-page volume published in 1974 by the Oxford University Press, the author, John R. Baker, Reader Emeritus in Cytology of Oxford University concludes his study by rating the human races on their innate capacity for originating civilization and ethical standards and finds the African blacks in last place. This assumption, whoever makes it, becomes entangled in all sorts of beliefs and activities—bringing light to the heathen, carrying the white man's burden, dealing with "lesser breeds without the law," bringing aid to undeveloped peoples, ensuring self-determination for South Vietnam.

Deplorable as these five complicating influences have been on the thinking and writing of scientific investigators of human variation, their effects on popular attitudes have been even more vicious. The idea of the Platonic type forms the basis for the belief that some physical trait such as coarse hair or a broad nose is proof that an individual represents some fancied type and consequently has a whole complex of other characters, behavioral as well as anatomical. The myth of pure race is used to denigrate anyone who happens to have some character—a wide skull or a third toe longer than the first—thought to have been absent from the elite founders of the stock. The confusion arising from the similarity of race and class makes possible the mutual reinforcement of racial and class prejudices; the emotional reactions aroused by the connection between race and sex make it easy for the individual to blame persons of other phenotypes or other groups for his frustrations and for his guilt. The up-from-the-brute concept of human evolution serves to justify a conviction of moral superiority over people who display what are assumed to be primitive characters—either anatomical or cultural. It facilitates the attribution of worthlessness and bestiality to other populations or their practices. Such a concept encourages American soldiers to regard foreign peoples with amused contempt, to classify Koreans and Vietnamese as "gooks," and to treat them as conveniently expendable.

UNITY AND VARIETY IN THE HUMAN SPECIES

HOMO SAPIENS A SINGLE SPECIES

If human races are analogous to animal subspecies and if a subspecies may develop into a full species, has this happened in the hominid line? Does man, in fact, have any relative who is to him what *Junco phaeonotus* is to *Junco hyemalis*? The answer is no. The Abominable Snowman is still in the same category of mythical creatures as the Loch Ness monster, and both seem likely to remain there permanently. Nowhere in the world are there two populations of manlike creatures living in close proximity for any length of time with no interbreeding. Wherever and whenever human populations have come together, interbreeding has always taken place. In applying the name *Homo sapiens* to all human populations, Linnaeus recognized that they constituted a single species, and scarcely any serious students have ever dissented since. In fact, no better example of the biological species than man could be found.

It does not follow, of course, that human mating takes place at random. Not only distance and physical barriers but also cultural practices break human populations into subgroups within which matings are more probable than they are across the subgroup lines. But the lines are never impassable. Time after time throughout history distinct social groups composed of people of different origins have come into existence within the same area as a result of conquest or migration; and often such groups have resisted assimilation and remained distinct for long periods. One of the best-known examples is the caste system of India, which began as the result of the conquest of that country by the Aryans, whose leaders became the Brahman caste. The system has had a long and complex history extending over 3,000 years. Although it has fallen into disarray and been officially abandoned in the present century, there is still evidence, in the frequencies of blood group alleles, for example, that some of the different social groups must have had different biological origins (Dobzhansky 1962, pp. 234–38). But in all such cases—the castes of India, the Jews, the black of the United States—there is also evidence that as time goes on, populations in the same area tend to become more alike. Human classes differ from ecological races of animals only in degree. Where they exist together, human classes resist genetic amalgamation with somewhat greater persistence than ecological races of animals. But in the long run the biological drives triumph over the prohibitions of mores, laws, and religion and over the inhibitions of the superego.

Nowhere within the human species is there any evidence for even the slightest tendency toward reproductive isolation. In other species of animals many cases are known where matings between individuals from populations that have long been separated show reduced fertility or produce offspring of less-than-normal fitness. If leopard frogs from Vermont are crossed with leopard frogs from Texas, most of the embryos die and only a few reach the tadpole stage. If gypsy moths from Europe are crossed with those from Japan, the resulting offspring include a substantial number of sterile intersexes. In both these cases the two long-separated populations have achieved genetic equilibria with genetic systems sufficiently different in informational content that when gametes from the two sources are combined, the resulting zygote is not piloted through a normal development.

Within some species of *Drosophila* a more subtle type of genetic incompatibility is known. Crosses between certain strains are fully fertile and even produce flies larger and more vigorous than the parents; but when these hybrids are interbred, their offspring, on the average, show less vigor and lower fecundity than their parents or even their grandparents. In these cases, the gametes from the different strains when paired at random produce vigorous individuals, but when these individuals form gametes, many of the new genetic combinations that result are unable to direct development with the effectiveness that characterized the gametes within the two different strains.

There is not the slightest evidence that any similar phenomenon occurs within the human species. In popular lore there is, of course, a luxuriant crop of old wives' tales and fatherless rumors defaming half-breeds and attributing to them all sorts of baleful characters. As a result of increasing human mobility during the last century and a half, matings between human beings drawn from widely distant and phenotypically unlike populations have occurred in large numbers in many places—in the Americas, in Africa, in the Pacific Islands. It is not possible to prove a universal negative by inductive procedure, but several extensive studies of hybrid populations have been made and they show no evidence for reduction in viability, fertility, or functional efficiency in the first, second, or later generations of hybrids between unlike human populations. These studies include one on Dutch-Hottentot hybrids in South Africa made more than sixty years ago (Fischer 1913), one made in the 1920s on the descendants of the mutineers of *The Bounty* (Shapiro 1929), and an elaborate study of 179,000 babies of mixed Caucasian-Oriental-Polynesian descent born in Hawaii between 1948 and 1958 (Morton 1962). Just as there appears to be no evidence for heterosis in the human species, neither is there any for hybrid disability. Both these phenomena would be likely if there were fundamental genetic divergence between human races.

Every biological species is unique. If it were not, it would not be a species. *Homo sapiens*, however, stands alone in the entire animal kingdom in two extremely important respects: the extreme plasticity of his potential behavior patterns, and his ability and need to develop within a group an integrated set of practices based on the use of language, abstract symbols, and tools—in short, culture in the anthropological sense. These peculiarities of the human species are probably related to the absence of any tendency toward reproductive isolation between populations and the lack of any evidence for hybrid disability.

The diagnostic characters that put man as a species in a very special category arise from the paradox that man has made a specialty of being unspecialized. Instead of growing thick fur against the cold, man made himself clothing; instead of developing flippers for swimming, he made a canoe and paddles. Through his use of tools, symbols, language, techniques, and social organization, man adapts his environment to himself rather than himself to his environment. By keeping himself versatile and plastic and altering the things around him for his own tastes and convenience, man has been able to flourish in a greater variety of environments than any other animal. In fact, this modification of the environment has gone so fast and so far that man faces a present crisis as a result. If it proves impossible for him to make the required environmental changes without at the same time polluting the air, water, and soil and destroying the balance between other plants and animals, man's own adaptation to existence will be radically altered and, in all probability, catastrophically reduced.

A creature of such ecological versatility must, by nature, be very unlikely to leave any ecological room for closely related but distinct species. To survive in the same locality, distinct but closely related species must have slightly different but distinct ways of life; they must occupy slightly different ecological niches. But man spreads over many niches and apparently has been doing this for a very long time. Not many years ago most students of fossil man interpreted the record as evidence that several fossil groups—Neanderthal man, for example—represented species not on the direct line of descent of modern man. This meant that Neanderthal man and some other species that was destined to become present-day *Homo sapiens* must have been contemporaneous. During the last thirty years this interpretation has been generally abandoned. It seems most likely that Neanderthal man was merely one population in western Europe of the many other human populations living elsewhere at that time, that all of them belonged to a single species and that Neanderthal as a distinct group was swamped and absorbed into other groups.

Going farther back to the period from one to four million years ago, the question of how many species of hominid existed at the same time is less easy to answer. There is an extensive, complex, and rapidly growing fossil record from East Africa for this period. It is fragmentary and very difficult to interpret, but C. L. Brace (1979), formerly the leading protagonist of the theory of the single hominid line, and R. E. F. Leakey (Leakey and Walker 1976, Leakey and Lewin 1978, and Walker and Leakey 1978) are now in agreement that *Australopithecus robustus* and *Homo erectus* were contemporaries in the Koobi Fora region of East Africa about a million and a half years ago.

No fossils that can be dated later than one million years ago have been identified as *Australopithecus*; so there is little doubt that since that time the human lineage has been represented by a single species. This is what one would expect of an animal of such ecological versatility. This versatility is the result of man's culture and the arguments over when the hominid line became monospecific are related to the time at which the developing culture bestowed the overwhelming versatility. At any rate, there has been a single species of man for a very long time. Man has also been a wanderer and his versatility and his mobility go a long way toward explaining the relatively slight and superficial differentiation between contemporary human subspecies.

THE ADAPTIVENESS OF HUMAN VARIATION

In trying to arrive at an understanding of the adaptiveness of human variation, one must remember that variation is both individual and geographic. Within populations the polymorphism is substantial. Individuals differ in almost every conceivable character: in eye color; pigmentation of skin and hair; form, quantity, and distribution of hair; stature; body build; shape of nose; prognathism; relative length of trunk and limbs. What is the explanation for this phenotypic variety?

It is possible to argue that individual variation is necessary so that the various roles required by the social division of labor can be filled by persons who are well fitted for them and who therefore find them congenial. Among the social insects, different roles such as worker or soldier are performed by very different phenotypes produced during development by differences in rearing. It might be claimed that in human societies small, lightweight individuals are produced to fill such occupations as jockey or chimney sweep, tall men to be basketball players, and heavy muscular types to be stevedores. The very statement of this hypothesis reduces it to absurdity. Insect societies are rigid organizations with a few clearly differentiated classes, each with narrow potentialities. Human societies are vastly more complex with many more roles, and there is no one-to-one correspondence between phenotype and socioeconomic role. The majority of men and women are potentially capable of filling many roles; the individual destined because of his phenotype to a single occupation does not exist.

Another conceivable explanation for polymorphism is that all existing populations have resulted from the interbreeding of phenotypically different groups and that we have a continuous segregation of the different original elements. This is really another form of the pure race theory and there is little evidence for it. Wherever we look, human populations, like animal populations, are polymorphic. Populations consisting of a single phenotype do not exist. There is no material evidence that they ever did and no other reason for thinking so.

The individual variability within human populations is evidence of the enormous amount of genetic variability characteristic of all populations of sexually reproducing higher organisms. Practically all of the many variable characters are produced by polygenic systems. This means that many different genetic combinations can produce very nearly identical phenotypes. When the genes from two phenotypically similar individuals are recombined in different ways, as happens at gamete formations and fertilization, the new combinations give an array of phenotypes including not only some similar to the parents but also others of varying degrees of divergence.

The visible phenotypes reveal, of course, only part of the genetic variability. There are also hidden differences in metabolism and biochemistry, in internal structure, and in functional performance. That in a population at equilibrium a complex array of phenotypes is maintained generation after generation does not mean that all genotypes and all phenotypes are equally adaptive. Many extreme phenotypes are obviously ill-adapted; others, for subtler reasons many of which we do not understand, leave no offspring. Selective forces are erratic. Not all individuals are subjected to the same stresses or suffer the same vicissitudes. Huxley's inexorable hidden chess player is effective only in a statistical sense and over many generations. What keeps the array of phenotypes and the range of variability constant is that, given the gene pool of the population and the selective forces of the environment, the gametes produced by one generation will, when combined according to the prevailing mating system, produce essentially the same array of genotypes and phenotypes in the next. It is not primarily the adaptiveness of the individual phenotypes which preserves the variability of a population, but the effectiveness of the array of phenotypes of one generation in producing a similar array in the next.

To have produced geographic variation, however, some other forces must have been at work. Why should the relative frequencies of phenotypes change as we go from one area to another? The most reasonable explanation is that selective pressures differ in different places. It is, of course, impossible to refute this statement. It must be true. And yet, the number of cases is extremely small in which we can clearly correlate a high frequency of given hereditary character with the type of environment that, as we understand the character, should give it a selective advantage.

One character that appears at first consideration to be adaptive is the dark skin of the peoples of Africa south of the Sahara. When human skin is exposed to sunlight, if the part of the spectrum known as the far ultraviolet penetrates to the living cells of the *stratum Malpighium*, it can have three separate effects: (1) if intense enough, it can damage and even kill cells and result in a sunburn of greater or less severity; (2) it can stimulate the melanocytes in the *stratum Malpighium* to produce melanin, resulting in the darkening of the skin— tanning; (3) it can change the substance called 7-dehydrocholesterol into vitamin D. The last plays a part in calcium metabolism. Its deficiency produces rickets in children and bone abnormalities in adults. An excess of vitamin D causes calcium deposits in soft tissues, resulting in serious complications in the kidneys. It looks, therefore, as though the effect of natural selection is to keep enough pigment in the skin to prevent damage from ultraviolet radiation but not so much as to prevent the formation of vitamin D. Since the number of hours and the intensity of solar radiation vary in different parts of the world, selective pressure for heavy pigmentation should be high where the radiation is high and low where the radiation is low.

In general, it is true that populations near the equator tend to be more heavily pigmented than those in the far north. But when local populations are examined in detail, some very puzzling inconsistencies show up. Among the most heavily pigmented populations of the world are those stretching across central Africa from Senegal through Cameroons and the Congo region to Kenya. Yet many of these people live in the tropical rain forest, where the sunlight reaching the ground is much less intense than that in the savannah country to both the north and south. More curious still is the observation that the Hottentots and Bushmen of southwest Africa who inhabit desert country with extremely intense solar radiation are much lighter than the blacks of the rain forest. Furthermore, it is known that the Bushmen and Hottentots were pushed south during historic time by more heavily pigmented groups from the north. Since the Bushmen now live under conditions of very intense solar radiation, and since they formerly lived farther north where the intensity of radiation must have been even greater, it is difficult to understand why they are distinctly less pigmented than the Negroes.

In Asia, Indonesia, and Australia the situation with reference to pigmentation is even more confused. There are dark-skinned populations in southern Arabia, South India, New Guinea, Melanesia, and North Australia, but almost none of these is so heavily pigmented as those of equatorial Africa. As one goes north on the continent of Asia or south in Australia, pigmentation lessens. But in Sumatra, Java, Borneo, and Celebes, all of which lie within ten degrees of the equator, the populations are only moderately pigmented.

The least heavily pigmented populations of the world are those of northwestern Europe where the high latitude and frequent cloud cover make for low levels of solar radiation. This fits the adaptive hypothesis perfectly. In Scotland and Scandinavia one would not expect natural protection against sunburn to have high adaptive value. One would expect the ability to produce vitamin D with a minimum exposure to sunlight to be advantageous. But in northeast Asia, Alaska, and southern Chile, where the average intensity of solar radiation is similarly low, the native populations never became depigmented to the extent that the Picts and the Vikings did.

Many of these anomalies of the distribution of pigmentation can be explained away by postulating various migrations of peoples in prehistoric times. Loring Brace (1969) has worked out a plausible explanation for the modern distribution in this way. He argues that Negroes did not go into the rain forest until relatively recent times when the acquisition of iron tools and special agricultural techniques made it possible for them to live there. In India, he contends, the lighter-skinned Indo-Europeans migrating southward diluted the earlier dense pigmentation there. As for Indonesia, the former dark-skinned inhabitants, he says, were effectively replaced by paler Mongoloids who migrated from southeast Asia. These deductions may be perfectly valid. They require many and complicated migration patterns to support them. There is no doubt that much migration occurred. But there is so little direct evidence on the details of migrations that one is tempted to postulate those movements that explain the existing distribution of pigmentation. It is still a puzzle why the northern Australians are lighter than the central Africans. And there is no very convincing argument to make clear why there was no depigmentation of the population of northeastern Siberia comparable to what occurred in northwestern Europe.

The adaptive significance of the form of the nose is another subject on which there has been much argument. Some writers have claimed that a high, narrow nose is an adaptation to a cold climate, for such a nose guarantees that inspired air will have been warmed by the time it reaches the lungs. It is true that in northern Europe the modal phenotype is characterized by a high, narrow nose. But a very similar type of nose is also characteristic of the hot dry regions of southwestern Asia and northern Africa. In fact, the argument has also been made that a high, narrow nose serves to moisten desert air. An elaborate argument to prove that the flat rather than the high nose is an adaptation to cold was made by Coon, Garn, and Birdsell in 1950. They contended that the

Mongolian face had been evolved in Siberia during the Pleistocene when large areas of that region were unglaciated but were subjected to winters of extreme severity. The flat nose and the fold of skin on the upper eyelid were developed, they said, to lessen the likelihood of frostbite or freezing. But whether or not the Mongolian face is an adaptation to cold, it certainly has no selective disadvantage in the heat. Large populations of Mongolians have been living for thousands of years in the tropical and subtropical climates of southeast Asia. The correlation between nose form and climate is certainly not simple.

There are two basic reasons why one is likely to be led astray if one assumes relations between phenotype and environment. In the first place a character like pigmentation or nose form is not an independent unit that can be altered without regard to the rest of the organism. Pigmentation, for example, may be influenced not merely by solar radiation but by many anatomical and physiological factors, such as microstructure of the skin and the capillaries or the details of amino acid metabolism. The form of the nose is closely related to the whole facial structure and especially to the bones and muscles of the jaws. There are undoubtedly many different ways of combining the various related elements for solving a given problem in adaptation, and once a satisfactory equilibrium between the various polygenic systems has been found, genetic hemeostasis would tend to resist radical alterations of the modal phenotype even in the face of selective pressure.

In the second place, environmental influences are complex and fluctuating. We have already pointed out that man is more effective and more versatile in modifying his environment than any other animal. For this reason, as human technology has increased in complexity and effectiveness, the nature of the environmental influences has changed. A cold climate exerted one series of selective pressures on primitive hominids who had neither fire nor clothing, and quite a different series on populations who had learned to use fire and to clothe themselves. The effect of solar radiation on man is determined not merely by the number of hours of sunshine and its intensity but also by the amount of actual exposure. The latter is influenced by the pattern of activities of the members of a population and by the methods they employ for shielding themselves. An Australian aborigine naked in the desert is certainly subject to different selective pressure with respect to skin pigmentation than a Bedouin on the Sahara wearing a burnoose.

In all present-day human populations the modal phenotype is the result of the interaction of many polygenic systems built up in past time by fluctuating and changing environmental pressures and consolidated into a functioning whole under the influence of genetic homeostasis. What seems to us a striking character such as a black skin, a pronounced epicanthic fold, or peppercorn hair, however it originated, may no longer have any selective advantage.

The pitfalls inherent in making judgments about the adaptiveness of human characters are strikingly illustrated by recent discussions among medical geneticists. Tay-Sachs disease is a lethal condition produced by a single recessive autosomal gene. In the homozygous recessive, as a result of a defect in lipid metabolism, progressive degeneration of the central nervous system sets in soon after birth and death usually results before the end of the second year. Among the Ashkenazim, the Jews of central and eastern Europe and their descendants wherever found, the incidence of the disease is high, 158 cases in every million live births. Among the Sephardim, the Jews of the Mediterranean countries, the incidence is about one hundred times less, 1.7 per million live births, and the non-Jewish population of the United States has an incidence not demonstrably different from that of the Sephardim. Because of the very high incidence among the Ashkenazim the disease has been the object of much study and research. Since 1970 it has been possible to identify the heterozygous carriers by a simple test. Screening of Ashkenazic populations is now common practice and many of these doomed children are now identified *in utero* and aborted.

The most likely explanation for the very high incidence of the allele among the Ashkenazim is that the heterozygotes, who are normal and asymptomatic, must at some time have had a selective advantage over the homozygous normal. It can be shown mathematically that if the fitness of the homozygote is defined as 1.0, a fitness of 1.01295 for the heterozygote will result in an equilibrium producing 158 homozygous recessives per million live births. The Diaspora, the scattering of the Jews throughout the Western world, occurred about fifty generations ago. It is reasonable to assume that the separation of the Ashkenazim from the Sephardim was not made on the basis of the incidence of Tay-Sachs disease and that therefore the incidence must have risen among the Ashkenazim since that time. It is, of course, true that from the time of the Diaspora until the nineteenth century the Jews lived a segregated existence; and their environment could have been critically different from that of their neighbors in a way that might have given advantage to the carriers of one recessive Tay-Sachs allele. It is also quite conceivable that the Jewish way of life in the *Shtetlach* of eastern Europe may have given the heterozygotes an advantage, while the way of life under the very different ecological conditions of Cordova or Seville did not.

The increased fitness of 1.01295 for the heterozygote is sufficient to give an equilibrium with 158 cases per million, but if fifty generations ago the Ashkenazim had the same incidence as the Sephardim of today, this amount of increased fitness would not yet have led to the present Ashkenazic level. It would take 350 generations to reach 158 cases per million. To arrive at the present incidence in the Ashkenazim in fifty generations, the fitness value for the heterozygote would have to have been 1.05263. But with such an advantage the Ashkenazim would not now be at equilibrium; the incidence would still be rising and would come to equilibrium in another fifty generations with a final equilibrium incidence of 2,267 affected births per million. On the basis of these calculations it has actually been suggested that the incidence of Tay-Sachs disease is now increasing among the Ashkenazim (Myrianthopoulos and Aronson 1966, Shaw and Smith 1969).

Much of this argument is speculative, and the deductions are not susceptible to clear demonstration. It is hard to explain the observed facts without invoking the hypothesis of increased fitness of the heterozygotes. What is clearly not justified, however, is that the assumption that the heterozygous advantage had a constant value from 100 A.D. to the present. It is much more likely that the advantage varied with time and place. It seems to have had a greater effect in Poland, Lithuania, and Russia than in the Rhineland or the Balkans. And it is quite likely that with the decline in the difference between the Jewish and the non-Jewish way of life the advantage has declined also. It may no longer exist at all.

The most widely accepted illustration of allelic frequencies in human populations being influenced by a selective advantage of heterozygotes is afforded by the several deleterious alleles that have high frequencies in populations where malaria is or has been endemic (Livingstone 1967, Bodmer and Cavalli-Sforza 1976). One of the best known of such alleles is hemoglobin S. Hemoglobin molecules containing this mutant beta chain, which has only a single amino acid substitution, crystallize when the oxygen level is low and distort the red cells in which they are found into a sicklelike shape. Hence the name sickle-cell disease is applied to the abnormalities suffered by the persons homozygous for the allele. The distorted cells clog the capillaries and cause, periodically, serious and painful crises that put the affected individuals at a distinct selective disadvantage. The heterozygotes, who are said to have sickle-cell trait rather than disease, are much less severely affected and are essentially asymptomatic under normal conditions. These heterozygotes are less seriously affected by malaria than are individuals homozygous for normal hemoglobin and in parts of West and Central Africa the frequency of the hemoglobin S allele is above 10 percent. Although the mutant homozygotes have a reduced life expectancy and contribute fewer offspring than either the heterozygotes or the normal homozygotes, the heterozygotes, because of their resistance to malaria, have an advantage over normal homozygotes in both viability and fertility and this keeps the frequency of the mutant allele high.

There are some other mutant alleles that tend to be high where malaria is common. Some of these such as hemoglobin C and hemoglobin E, like hemoglobin S, affect the structure of hemoglobin chains. Others—the thalassemias—affect the ratio of alpha to beta chains. Still others affect the level of activity of the enzyme glucose-6-phosphate dehydrogenase (G6PD). There is a broad band of territory extending from the western Mediterranean and West Africa south of the Sahara eastward to Arabia, Iran, India, Southeast Asia, and Indonesia where *Plasmodium falciparum*, the most virulent of the malarial parasites for man, is present and where one or more of these alleles are found with frequencies well above those found elsewhere in the world.

There has been much argument as to how these mutant alleles provide resistance to the parasite. The evidence points for all of these alleles to a shortening of the life of the red cell, either in general or when it is parasitized. The malarial parasites, once established in an individual, proliferate within the red cells and when mature, escape from the cell they have consumed, to invade, individually, fresh unparasitized red cells in which they repeat the cycle. The developing parasites will not be able to build up a dense population if the cells they invade are fragile and die or disintegrate before the parasites mature. Recent work (Friedman and Trager, 1981) has demonstrated that the parasites can grow only in a red cell with an intact membrane that keeps the concentration of potassium within the cell well above that in the surrounding serum. Cells containing the abnormal hemoglobins develop damaged and leaky membranes, especially when infested by parasites, and the resulting fall in potassium concentration is fatal to both cell and parasites. The enzyme G6PD produces a substance necessary to keep the red cell membrane in good repair and cells with an abnormally low level of enzyme activity also develop leaky membranes under stress from the presence of parasites.

Although experimental infection of normal individuals and of heterozygotes for hemoglobin S in 1953 (Allison 1954) showed a pronounced difference in resistance to *falciparum* malaria, it has not been possible to demonstrate clear-cut differences in all populations for all the alleles, nor have any quantitative measurements of selective advantage been made. Nevertheless, the overall correspondence between allelic frequencies and present or past incidence of malaria is overwhelming evidence that in all these cases heterozygous advantage has increased the frequency of deleterious mutants.

It is ironic that malaria and tendency to anemia should provide the only authenticated examples of single gene heterosis. From an evolutionary point of view these cases are probably abnormal and transient. The effects, both of the mutants in homozygous condition and of the parasites in normal individuals, are extreme. Both parasite and host would do better to work out a less virulent interaction. There is reason to believe that this violent relationship is the result of a recent collision. It is quite likely that malarial infection was rare in human history before populations settled down to practice agriculture. This necessitated clearing land in areas with frequent rainfall and this provided puddles in which the mosquito vectors of malaria could flourish. This would mean that the interaction has been going on for only a few thousand years, not long enough for the parasite to have developed means of exploiting its host without threatening the host's existence or for the host to have found less painful means of holding the parasite within tolerable bounds. This kind of heterozygous advantage is violent and obvious and provides a textbook illustration. The sort of heterozygous advantage that contributes to the stability of the modal phenotype under conditions of great genetic variability is of necessity more subtle and more difficult to demonstrate in detail.

Questions of fitness and adaptiveness are fascinating and challenging, but they are difficult and allusive. They are particularly treacherous when applied to human populations. The long human generation time makes the collection of critical data very difficult, and the emotional concomitants of a deduction may be intense. It is one thing to prophesy that pink-banded snails are likely to increase in frequency if the number of thrushes who eat them declines and something very different to tell the Jewish population of Bridgeport, Connecticut, that it can look forward to a steady increase in the incidence of Tay-Sachs disease over the next 1,800 years.

RACIAL HISTORY

As long as people assumed that there had formerly been pure races of men, racial history was a description of the way in which these pure strains had been crossed to produce the variable, mongrelized populations of today. There has never been any evidence for the former existence of pure races, and a racial history based on this premise could not be more than fanciful speculation.

From what we know of the variability of human and animal populations, if racial history means anything, it is an account of the different variable populations of man which have existed at different times in different places, the traits that have characterized their modal phenotypes, how much interbreeding there has been among them, what their relative rates of increase have been, and how they have been related to each other culturally. It would be an account of the ebb and flow of variability, both individual and geographic, within the human species.

Tracing such a kaleidoscopic process is not easy. In the case of the frequency of Tay-Sachs disease among the Ashkenazim, about whom we have considerable historical data, it is impossible to say just how, in what century, and in what places that incidence increased, what the rate of increase was at any given time, and whether the incidence is now still increasing. Within the last three decades several attempts have been made to estimate the percentage of the genetic material in American Negroes which can be traced to white ancestors. The estimates are made by comparing the present frequencies of certain alleles among blacks with the frequencies of the same alleles among the populations of

West Africa from which the slaves were imported into the United States. No allele is known which is totally absent in West Africa and has a frequency of 100 percent among American whites. One of the alleles of the Duffy blood group system comes closest to this ideal. It has a frequency of about 43 percent in American whites and a very low frequency—less than 5 percent—among west and central Africans. Estimates made on the basis of this allele vary from 22 percent alleles from white ancestors in the blacks of Oakland, California, to 4 percent in a black population in Charleston, South Carolina. Using data on other alleles and other populations, estimates of the proportion of genetic material from white ancestry in American blacks range all the way from a few per cent to more than 50 percent (T. E. Reed 1969). The important point is that where we have good historical knowledge and considerable information on allelic frequencies, there remain formidable statistical problems and gaps in our knowledge of the adaptive values of different alleles. These make it almost impossible to give a precise figure for the genetic relationship between two populations.

As we go farther back in historic time the difficulties of reconstructing racial changes in terms of the amount of interchange of genetic material between populations becomes greater and greater; and before recorded history the problem becomes Herculean. We have little notion of how much the invading Visigoths or the later invading Moros altered the genetic makeup of the population of Spain. We know next to nothing of the genetic differences or similarities among the populations of pre-Roman Italy. The same is true for the Greeks of the Homeric period.

As soon as we get back to a period with no written record our knowledge of previous human populations has to be based on archeological remains. These include—in addition to human bones—tools, ornaments, and refuse. All of these can provide considerable information about the kind of life a people led, but only the bones can give us any direct evidence of their physical characteristics. From skeletons we can deduce the overall size and general body build, head form, and relative limb lengths, but they are uncommunicative on all those characters that are built into the soft parts and which are generally much more important in determining our emotional reactions to people. Skin color; the color, form, and distribution of the hair; the form of the features; and the aspect of the face—of these characters we know absolutely nothing for any prehistoric men from *Homo erectus* of Java through Neanderthal man, Cro-Magnon man, and the various later groups who left bones and artifacts here and there throughout the world. The archeological remains found in Europe are the most extensive and the most extensively studied. It can be shown that in some localities there have been changes through time of skeletal characters—from large heads to smaller, from long heads to rounder ones—but just what these changes meant in terms of genetic difference is far from clear and what sort of external physical features characterized the various peoples is wholly unknown.

Even though archeology cannot give us the story of genetic differences and changes of the past, it might be possible to deduce from the present distribution of physical characters the former distributions and the changes that they must have undergone to produce the present state of affairs. If we knew nothing of human history, observations on the physical characters, social organization, and linguistic habits of the people of the United States would certainly lead us to conclude that in addition to the predominant white English-speaking population that had recently been assimilating large groups of non-English-speaking immigrants, there had been a substantial black population in the South which had been gradually diffusing north and west, another population in the Southwest speaking Spanish and diffusing less rapidly, and some small, scattered, relict populations here and there, the larger and more persistent of these also in the Southwest. But the events responsible for the present arrangement of population groups have occurred only within the last three centuries, and some former populations have disappeared without trace. We would never suspect that three hundred years ago the population of the Midwest was Indian and New York was a Dutch city.

The wealth of information concerning physical and chemical characters that a living population can give us is not very enlightening on the question of how the population came to be that way, and the farther back we try to trace the ebb and flow of variation, the more difficult the task becomes. It is simply not possible to reconstruct the history of the interrelationships of interpopulational variation from the present distribution of characters.

As we have seen in our discussion of the adaptiveness of skin pigmentation, the present distribution is replete with inconsistencies and anomalies. There is a general tendency for skin color to be darker near the equator, lighter to the north and to the south. But the patchwork distribution of skin color over the world defies any simple historical explanation. Various plausible hypotheses of migration and interbreeding can be constructed to explain it. But all of these depend on what we assume to have been the earlier situation, and the farther back we attempt to go, the more different possible situations can be made consistent with the present facts.

What is true of skin pigmentation also holds for hair form. Human hair varies from fine to coarse and from straight through wavy to coiled. The extremes are to be found in the coarse straight hair of the Mongolians and in the fine hair of the Bushmen, which coils so tightly into intertwined bundles that it leaves bare patches of scalp between. Wavy hair is found among Europeans, Australian aborigines, and Polynesians. Hair coiled with different grades of intensity is found among African Negroes, Negrito groups in India and the Andaman Islands, on the Malay Peninsula and in the Philippines, and among Micronesians on many of the islands of the southwestern Pacific. There are all sorts of intergrades between the different hair forms of different populations, and within many populations there is great variability. Among Europeans hair form varies from straight through wavy to various degrees of curly, some of which approach coiled hair. Hair form, like skin pigmentation, is certainly a polygenic character, for it occurs in various grades and does not segregate as if it were a unit character. But several pedigrees are known—one of them among Norwegians—in which coiled or "wooly" hair was inherited as a single autosomal dominant.

If one takes some other character—nose form, lips, head shape, body build, it makes little difference which—one has no better luck in deducing from its present geographical distribution what the earlier situation must have been and how the present came about.

With the discovery of the mode of inheritance of red cell antigens such as the ABO blood group substances discussed in chapter 5, there was for a time great optimism that these characters would make it possible to unravel human racial history. These were simple biochemical traits, and single alleles producing them could be fairly easily traced in human pedigrees. The various alleles seemed to be adaptively neutral, and it looked as though a study of the geographical distribution of the frequencies of such alleles would show just how the assumed early simpler arrangement of human populations had been disturbed and jumbled to produce what we observe at present.

In 1950 William Boyd published *Genetics and the Races of Man*. This was an attempt to use blood group antigens as a means of classifying human races. It was a very careful and competent marshaling of the evidence, but it led to a classification recognizing five races: European, Asiatic, African, American (Indian), and Australian. In addition it suggested a former "early European," of which the Basques were a remnant, to account for the high frequency of Rh-negative alleles in western Europe. The whole classification was not very different from the traditional ones based largely on skin color. There are differences between different populations in the frequencies of the alleles in the various blood group systems. There is no need to go into them in detail here. These differences, however, are quite as complex in their distributions as the visible characters are, and they demarcate differences between human races no more sharply. There is now much less optimism among geneticists and anthropologists that biochemical characters are going to resolve questions of genetic history than there was in 1950.

The real reason why racial history cannot be read from present distributions of characters is that there are two fundamental ways of producing a change in the characters of a population: a large number of immigrants with different characters may come in, or selection within a population may reduce the frequency of certain characters and increase the frequency of others. If we know only that the frequencies of characters have changed, there is no way to be sure which process produced the change. In Europe there is a cline in the frequency of B blood group alleles from west to east. The frequency is very low along the Atlantic coast and increases as one goes toward eastern Europe. It is less than 10 percent in Spain, France, and Britain and more than 25 percent along the north shore of the Caspian. There is no doubt that for many thousands of years there have been movements of population from eastern Europe toward the west. Are we to explain the low frequency of B alleles in western Europe as a result of migration from the east which brought B alleles into a population that formerly had even fewer? Or is it higher selection pressure against the B allele in the west that keeps its frequency there below what it is in the east? Either explanation could account for the cline. We do have evidence that the alleles for blood group antigens are not entirely neutral. Certain genotypes in the ABO system seem to be more susceptible to certain diseases—peptic ulcer and carcinoma of the stomach. The differences in susceptibility are not great, and the diseases rarely develop until the victims have reproduced. But if these differential susceptibilities exist, there may be other selective pressures that we have not yet uncovered. There is good evidence that the alleles of the different blood group systems interact. Mother-child incompatibility in the ABO system, for example, reduces the likelihood that incompatibility in the Rh system will result in hemolytic disease of the newborn. One can certainly not regard the blood group antigens as convenient neutral markers for tracing ancestral wanderings.

In cases where we find two or more different populations that are phenotypically similar, does this mean that they had a common similar ancestry or that selective pressures in similar environments have produced similar combinations of characters? This kind of development has occurred many times in geological history even among species of remotely different animals. Sharks, bony fishes, aquatic reptiles known as ichthyosaurs, and porpoises have developed general body builds of astonishing superficial similarity as a result of their common marine environment, but their similarities are the result of similar selective pressures, not of common descent.

Among humans there are populations of short, dark-skinned people with coiled hair found in widely separated places: the Pygmies of the Congo area and Negrito populations in India, Malaya, the Andaman Islands, and the Philippines. One can postulate that all these groups were once one. They may have migrated from one point of origin to their present widely scattered locations. They may once have had a continuous range over the intervening areas in which they have been more recently replaced by other migrating people. Or these different dwarfed populations each may have arisen in its own habitat by converging in phenotype as a result of similar selective pressures. In general, these populations are found in hot, wet, forested areas. On the basis of present phenotypes and distributions it is impossible to say which of these processes provides the true explanation.

When migration of a human population takes place, there is almost invariably a cultural as well as a biological element involved. If the immigrants are culturally indistinguishable from the host population, the two will probably merge rapidly and the migration will have little historical consequence. Something approximating this process occurred in western Europe and in some parts of the United States, particularly in the South, during the nineteenth century when many cities grew in population as a result of an influx from the surrounding countryside. In other cases, of course, especially in the North of the United States, cities grew as a result of immigrants from far away who were culturally and often genetically distinct from the earlier urban population. In these latter cases, merging of the earlier population and the later immigrants has been slower. Sometimes culture migrates without any substantial shift in population at all. One of the most dramatic examples is that of modern Japan. In the past century Japan has taken over and assimilated Western science, technology, and industrial organization. This has occurred with no substantial genetic contribution from the West.

Migration may occur at the lower socioeconomic levels, as was the case with most of the individual nineteenth-century migrants to the United States. The volume of this type of migration has become much greater in the last two centuries than it ever was before, but it was certainly not unknown in the ancient world. Migration at the upper socioeconomic levels usually takes place in the form of domination either through military conquest or by means of political or economic control. Conquest is almost certain to be accompanied by a cultural difference between conquerors and conquered, but its genetic effect on the conquered population has varied from virtual annihilation to almost no change at all. Of the Spanish conquests of the sixteenth century, the one in Costa Rica resulted in an almost complete replacement of the indigenous people by a Spanish population and culture. In Mexico a substantial Spanish element was injected into the population, but the various native peoples were not annihilated, and a genetic and cultural amalgamation is still working toward equilibrium. In the Philippines the genetic contribution of the Spanish was very slight; and in the three quarters of a century since Spanish control ceased, Spanish cultural influence has become steadily less evident.

Whether an invading population will dilute, swamp, or displace the host population depends on many things; but a host population is likely to be hard pressed if it is sparse and if its ability to utilize the resources of the territory is technically inferior to the invader's. The peoples who have fared worst as a result of European expansion in the last four centuries have been those whose hunting-and-gathering economy made the use of their resources very inefficient by European standards.

As we go farther back in history most of the migration of which we have knowledge had a strong cultural element, and we have more precise knowledge of cultural changes than we do of changes in allelic frequencies. The Norman conquest of England produced profound political and cultural changes, but genetic change of the population was probably slight. The Germanic invasions of the fifth and sixth centuries produced political and cultural change and probably involved substantial movements of population, but the degree to which they resulted in genetic change in the populations of western Europe is anybody's guess.

In prehistoric times there were movements of populations and there were cultural changes, but just how great the genetic changes were and to what degree they followed the cultural changes are not at all clear. The only evidence for genetic change is what can be read from fossil bones. Such remains are scarce—much less frequent than cultural artifacts—and in general they have been interpreted with an emphasis on differences rather than with an awareness of the great variability of the species. Undoubtedly, the different kinds of interactions between populations and cultures which history records were going on also in prehistoric time. There were probably cases of annihilation of one population by another, instances of different degrees of cultural and biological fusion, and examples of temporary conquest without long-term cultural or genetic effect. There were probably cases of the spread of culture without migration of people. It seems quite likely that the spread of Neolithic agriculture westward across Europe did not occur as one massive migration in which a new population completely supplanted an older one. Colonists did move from east to west, but there seems also to have been adoption of the new techniques by hunting-and-gathering people living in contact with settled farmers. It is also very likely that the spread was not a smooth process. There were undoubtedly friction and violence from time to time here and there, and some populations did supplant others.

The study of prehistoric man and of hominid evolution is fascinating and challenging. New fossil material is constantly being discovered, and new and better techniques are being developed to help in its interpretation. We will learn more and more of the story as time goes on. But we will probably never be able to trace back the origins of our present populations to the extent that we will be able to say, Populations A and B have 60 percent common ancestry but population C has 20 percent in common with A and 50 percent in common with

B. Neither is it likely that we will ever be able to say where the wing bars in *Junco hyemalis aikeni* came from. One may hope that we learn to stop looking for heraldic pedigrees to relieve our insecurities and bolster our prejudices and instead attempt to understand as completely as possible the complex flux of human variation.

THE SOCIAL MEANING OF HUMAN VARIATION

The present descendants of the seventeenth- and eighteenth-century immigrants into New England probably suffer as little from racial discrimination as any group in the United States. Yet it is impossible to delimit the group on the basis of phenotype. The variability is enormous. The hair may be tow, yellow, sandy, brown, or black. It may be fine or coarse, straight, wavy, or curly. Beard may be heavy or sparse, tough or light. Body build may vary from the Don Quixote to the Sancho Panza type. The eyes may be blue, gray, hazel, brown. One could go on and on listing the phenotypic differences. Within the group, however, there is no consistent discrimination against those members having some specific character or group of characters. There may be some deformed or pathologically unattractive individuals who, as a consequence, find themselves at a disadvantage; but those individuals are not regarded as a group to be kept within bounds. Within the established cultural group, human variation is accepted as an inevitable part of the nature of things.

It is usually when phenotypic difference becomes identified with a difference of cultural origin that is is likely to give rise to discrimination and become a divisive influence in human society. In nineteenth-century New England the first immigrants to establish themselves as a large, separate group were the Irish, who began arriving in the 1840s. The phenotypic differences between the Irish and the older New Englanders were subtle and would probably not have been apparent to a Japanese or an Ethiopian. But the cultural differences were sharp. The newcomers were poor and had to take their place in the lowest socioeconomic stratum; they spoke with an accent—a brogue—and their Roman Catholic religion raised questions in the minds of the older inhabitants concerning their moral, social, and political worth. Between the two groups a bitterness and mutual contempt developed, and vicious discrimination against the Irish remained a fact of New England life until well into the present century. Phenotypic characters such as red hair, freckles, or "the map of Ireland on the face," which were not common to all the Irish nor entirely absent from the New Englanders, came to be regarded as grounds for suspicion in determining the acceptability of an individual.

But within less than a century and a half since their first arrival, this formerly despised group has been pretty effectively assimilated into the general population; has produced many illustrious persons in the arts, the professions, business, and politics; and has even contributed a president of the United States. Physical stigmata of Irishness have simply ceased to be a factor in evaluating an individual.

The history of blacks in the United States has been very different from that of the Irish. Brought to the country in servitude rather than in mere poverty, kept in servitude until a century and a quarter ago, and in political, economic, and social subjugation since, the Negro has been treated by the dominant majority as unassimilable and the cultural gap between black and white has remained wide and deep. The phenotypic differences between blacks and whites are, of course, much greater than those between Yankee and Irish, but the importance of culture and former status in determining the limits of assimilation is shown when many persons indistinguishable from whites are kept in the inferior status of the black group because their ancestors include former slaves. In these cases, it is the cultural history and neither phenotype nor biology that is the determinant.

The intricate way in which cultural traits and phenotype are used to reinforce each other in classifying individuals can be observed wherever people make such judgments. Literature is full of illustrations. There is an interesting exchange in Galsworthy's *To Let*.

> "Hm," said Soames. "What does that chap Profond do in England?"
> Annette raised the eyebrows she had just finished.
> "He yachts."
> "Ah!" said Soames; "He's a sleepy chap."
> "Sometimes," answered Annette, and her face had a sort of quiet enjoyment. "But sometimes very amusing."
> "He's got a touch of the tar brush about him."
> Annette stretched herself.
> "Tar brush?" she said, "What is that? His mother was *Arménienne*."
> "That's it, then," muttered Soames. "Does he know anything about pictures?"
> (Galsworthy 1921, pp. 28–29).

Doubts about the man's credentials as a gentleman cause Soames to suspect his ancestry, and finding that his mother was Armenian had as calming an effect as realizing that the noise one has just heard was merely the straightening of a recently crumpled piece of paper. The irony of this illustration is underlined by the knowledge that the Armenians come from very near the Caucusus where Blumenbach's "most beautiful race" was found.

In George Eliot's *Romola* (1910) there is a scene (p. 209) where the heroine is talking to Lillo, her dead husband's illigitimate son by his peasant mistress. The author observes, "Lillo was a handsome lad, but his features were turning out to be more massive and less regular than his father's. The blood of the Tuscan peasant was in his veins."

In *Counseling in Medical Genetics* by Sheldon Reed, published in 1963, there is a list of tests to which an infant of questionable biological parentage available for adoption can be subjected to decide whether he will be able, when mature, to pass as white. This is recognition that certain arbitrary stigmata have become accepted as diagnostic of membership in a group with inferior status. In the absence of these stigmata, unrevealed descent from the inferior group is of little consequence.

What constitutes a race is a matter of social definition. Whatever a group accepts as part of itself is within the pale; what it rejects is outside. Acceptance and rejection are not absolute but can exist in various degrees. In the last three decades the racial situation in the United States has changed considerably and at present is in a state of flux. It is, however, interesting to compare the American racial practices of the first half of the present century with those of the Spanish Caribbean of the same period. In the United States the line of demarcation between black and white was traditionally extremely rigid. Anyone with any black ancestry was classified as black, and any white marrying a black (where this was legally possible) was written off by his family. In consequence, family bonds between blacks and whites were nonexistent, and casual social contact between the two groups as equals extremely rare. In Puerto Rico and Cuba a different situation obtained. The economic and social establishment was white and the majority of blacks lived in poverty. There was antiblack snobbery, and blacks were not members of exclusive clubs. But the line of demarcation was much less rigid than in the United States. Some black ancestry did not of itself exclude one from all participation in social, business, or professional life. It was not unheard of for whites to have aunts, uncles, or cousins who had some black ancestry and did not conceal it. Whites could have black friends and did not lose caste for entertaining them.

Historically, race in the United States has not been the same social phenomenon that it was in the Spanish Caribbean. In spite of the changes in the United States in the last three decades, the legacy of the old rigidity is still a reality. The black native of the United States sees the question of identity differently from the Puerto Rican or Cuban of African ancestry, because in the United States the dominant culture has defined and reacted to blackness differently than the dominant culture of the Spanish Caribbean.

When we look around the world we find that different cultures define race differently and that every culture has its own peculiar way of reacting to it. In Europe where the individual variation is very nearly as great as the geographical, class and linguistic differences have been the sources of social tensions. Class and linguistic groups have been thought of and spoken of as races, although the genetic cleavage along these lines is very slight.

Discrimination against the Jews, which has had a long and dismal history punctuated with outbreaks of violent persecution, has been directed primarily against groups having separate religious, cultural, and linguistic identity. Genetic considerations have been of little importance. As to both phenotype and biochemical characters, the Jews of Europe show convergence toward their non-Jewish neighbors, and in spite of folklore to the contrary, there is no recognizable Jewish phenotype. Where Jews have been culturally assimilated, they have often merged completely with other populations. It is significant that in the heyday of the British Empire the most imperial of its prime ministers was of Jewish origin. The holocaust of Nazi persecution was exceptional in that it pretended to have a scientific genetic basis. But without the long history of Jewish-Gentile antagonism, it is doubtful that this self-styled "scientific" quackery could have produced the horrors that were realized. It is probably true that the appearance in the Weimar Republic of culturally alien groups of Orthodox Jews as displaced persons and immigrants from eastern Europe and the Balkans served to make Nazi anti-Semitism more attractive to the general population, already discontented and frightened by the uncertain prospects of a world in flux.

In Europe generally, individuals with non-European phenotypes have been subjected to systematic discrimination only in cases where they have appeared as compact groups with different cultural practices. In London and other English cities where substantial groups of West Indians or Pakistanis are concentrated in certain neighborhoods where they follow their own separate way of life, they have aroused intense hostility on the part of their native English neighbors, and the question of racial prejudice has become a troublesome one in British politics.

In East Africa, people of Indian descent whose ancestors came there as immigrants two or even three generations ago and who in many areas constituted a distinct cultural and economic group of traders and shopkeepers have in recent decades been subjected to political and economic discrimination by the dominant black populations. In India, where phenotypic variability is very great by European standards, status has for thousands of years been determined by a complex system of hereditary castes. In the present century this old system has been substantially modified, but traditionally the castes determined in the most rigid fashion one's social and economic position and as rigidly limited one's aspirations. The castes, though hereditary, were neither phenotypically nor genotypically homogeneous internally. Nor were they clearly separable from each other in phenotype or in genotype.

All these illustrations emphasize that both what constitutes a race and how one recognizes a racial difference are culturally determined. Whether two individuals regard themselves as of the same or of different races depends not on the degree of similarity of their genetic material but on whether history, tradition, and personal training and experiences have brought them to regard themselves as belonging to the same group or to different groups. Since all human beings are of one species and since all populations tend to merge when they exist in contact, group differentiation will be based on cultural behavior and not on genetic difference. True, there is great geographical variation within the species but even at its most extreme there is no biological barrier to interbreeding and population fusion. In these circumstances, whatever classification of human beings is to be made will be determined by the cultural practices of the classifiers. This is true for sophisticated scientific classifiers. What races they recognize is determined by the aim of their classification and the convenience of applying it. In 1795 Blumenbach recognized five races. In 1962 Coon recognized five, although they were not exactly the same. In 1950 Coon, Garn, and Birdsell had divided human populations into thirty races. In 1962 Dobzhansky found it "convenient" to list thirty-four. Since there are no objective boundaries to set off one subspecies from another like the objective boundaries between different species, the manner in which one classifies human populations will be determined by one's tastes and predilections and by the purpose of the classification. It is worth noting that most of the classifying of human races has been done by Europeans or their recently displaced descendants. Hence the classifications have been made from the European point of view. Suppose that the developments in science and technology which occurred in Europe in the sixteenth century and gave rise to the Industrial Revolution and to European expansion had taken place instead in the Ganges valley. It is interesting to try to imagine what effect this would have had on the classification of human populations. Certainly, one of the primary races would be the Indian race, and the peoples of Europe would probably be regarded as a somewhat aberrant branch.

Although in different parts of the world the human species has developed populations so different in appearance as the Eskimo, the Irish, the Balinese, and the Watusi, these widely separated populations have diverged genetically to a remarkably mild degree. Human migration and interbreeding have maintained a continuous gene flow throughout the whole human population sufficient to prevent any really fundamental genetic divergence. As pointed out earlier, a deliberate search for a single hereditary biochemical character capable of distinguishing clearly between American white and west African blacks has not been successful. The great majority of human alleles are shared by all large population groups. Between groups the frequencies of a given allele may vary sharply, but complete absence of an allele in any large population is a rarity.

With the accumulation of data on alleles capable of biochemical identification—the same as those used by the molecular evolutionists to compare the chimpanzee with the human—various attempts have been made to measure the proportion of difference between individuals within populations and that between populations. Among others, Lewontin (1972) and Nei and Roychoudhury (1974) have wrestled with this problem. Different measures have been worked out which involve the application of fairly sophisticated mathematical analysis. The upshot of all this is that the methods agree in ascribing more than four-fifths of the genetic differences to those found between individuals and less than one-fifth to those found between major geographical areas—the human subspecies. This whole topic in all its complexity is summarized by B. D. H. Latter (1980).

Genetic variability *within* populations is greater than the variability between them. Within populations the phenotypic variability tends to be more continuous or graded. Between populations phenotypic differences are sometimes sharp but this is the result of different proportions and combinations, not of differences in the presence or absence of alleles. Selection within populations, while differing somewhat from place to place, has apparently been more powerful in the direction of producing plastic, educable individuals than of creating phenotypes finely adjusted to the climate, diet, and terrain of their place of residence.

It is true that many historical cases of bitterness, conflict, and bloodshed can be cited between human groups dissimilar phenotypically: blacks and whites in the United States, Afrikaners and blacks in South Africa, Chinese and Malayans in Malaya—the dismal list goes on and on. But it is not difficult to produce an even longer list of cases of suspicion, conflict, cruelty, and bloodshed between human groups having no significant phenotypic differences between them. Neither the Protestants of Ireland nor the Moslems of India, converts during historic time, were recruited on the basis of phenotype, but their relations with their neighbors—Catholics and Hindus, respectively—are often just as violent as those between racial antagonists. Repeatedly in human history landlords and peasants have been embroiled in internecine conflict—in Reformation Germany, in China during the Tai-Ping Rebellion in the 1860s, in revolutionary France. In recent years Communists and non-Communists have slaughtered each other in Indonesia; in Nigeria, Ibos and northerners have clashed; in Colombia, Liberals and Conservatives. Phenotypic differences are by no means necessary for the development of human tension, strife, cruelty, and slaughter.

Culture plays the major role in determining how human beings judge and react to one another. It is culture, not genotype, that leads one population to attempt to kill off another that it considers racially different. The purpose of this book has been to show that in spite of all their differences—genotypic because of the wide variety of alleles in the human gene pool, phenotypic because of the different interactions between genotype and environment, geographic because allelic frequencies vary from place to place, and cultural because the human biotic program makes culture a necessity but does not define its content— human beings constitute one species whose most precious asset is, in fact, their diversity. This diversity is directly responsible for the human species' resiliency in the face of continual environmental flux. To some extent man shares this resiliency with other animal species. But man's genetic variability includes in addition the most complicated of all polygenic systems, one that guarantees the cultural competence of the individual. This character man shares with no other animal and it gives him the awesome potential that he has. It is our hope that as more of us come to understand the biological unity of man, more of our energy will be diverted from intraspecies conflict to the promotion of human fulfillment.

GLOSSARY

ALLELE One of two or more different forms in which a gene can exist.

AMINO ACID One of the small organic molecules of which proteins are composed (see fig. 3).

CHROMOSOME A discretely packaged portion of the genetic complement consisting of DNA containing the genetic information plus several kinds of protein and some RNA.

DIPLOID Having the genetic information present in two versions, one inherited from each parent. The cells of most higher animals, including man, are diploid.

DNA Deoxyribose nucleic acid. The substance in which the genetic information is coded.

DOMINANT ALLELE A form of a gene which expresses itself in observable characters when it occurs in a single version of the genetic information (see Heterozygous).

EUKARYOTES All those organisms in which the DNA is confined within a nucleus. Among these are certain one-celled creatures such as yeast and the amoeba and all multicellular plants and animals, including man.

FLANKING SEQUENCES See Sequences.

GAMETES The reproductive cells: eggs and sperm.

GENE A functional unit of information in the DNA.

GENE POOL The aggregate of all alleles existing at a given time in all members of a breeding population. The genotypes of the next generation will be drawn from this gene pool.

GENETIC COMPLEMENT The total amount of genetic information coded in DNA necessary for the normal development and functioning of an individual. In most higher animals, including man, the genetic complement contains two versions of the genetic information, one derived from each parent and is said to be diploid.

GENOTYPE The genetic information coded in the DNA of an individual. The genotype is not directly observable. That part that is expressed may be inferred from observable characters; what is not expressed must be deduced from ancestry or descendants.

HAPLOID Having the genetic information present in a single version. In practically all vertebrates, including man, only the eggs and sperm are haploid; the body cells are diploid.

HETEROZYGOUS Said of an individual having two different allelic forms at a given locus in the two versions of his genetic information, having inherited a different allele from each parent.

HISTONE A special kind of protein associated with the DNA in the chromosome. Several different histone molecules form a core around which the DNA is wound to form a nucleosome. Another type of histone molecule is associated with the unwound section of DNA that links adjacent nucleosomes together.

HOMOZYGOUS Said of an individual having the same allelic form at a given locus in both versions of his genetic information, having inherited the same allelic form from both parents.

INTERVENING SEQUENCES See Sequences.

LOCUS The position in the linear extension of the DNA of a chromosome where a given functional unit is found. The term is used to refer to the location with relation to other functional units or to the functional unit situated there.

NUCLEOSOME The unit of packaging of DNA in the chromosome. It consists of a roughly spherical core composed of histones around which about 200 base pairs of DNA are wound twice. The separate nucleosomes are linked by about 50 base pairs of continuous DNA between each two nucleosomes. There are hundreds of thousands of nucleosomes in each chromosome.

NUCLEOTIDES Organic molecules that can be joined (polymerized) to form the larger molecules of DNA and RNA. All nucleotides consist of three parts: a phosphate group, a sugar, and a complex basic group containing nitrogen. In the nucleotides that form DNA the sugar is deoxyribose and the base may be adenine (A), guanine (G), cytosine (C), or thymine (T). In the nucleotides that form RNA, the sugar is ribose and the base may be any of the above except thymine, which is replaced by uracil (U).

PEPTIDE A molecule composed of a series of amino acids joined together by peptide bonds (see fig. 3).

PEPTIDE BOND The connection between two amino acids that form part of a peptide (see fig. 3).

PHENOCOPY An individual animal that because of some influence during development expresses a character not usually associated with his genotype but characteristic of some other genotype.

PHENOTYPE The observable character or characters of an individual. The terms is used in two senses: (1) to classify individuals with respect to a single character—blue eyes or O blood type, for example; (2) to refer to the whole complex of characters possessed by an individual.

POLYGENIC Said of a character, usually one with a continuous distribution like height, the expression of which depends on the interaction among the alleles present at several different loci.

POLYPEPTIDE A long peptide chain containing many—a hundred or more—amino acids.

PROKARYOTES Those organisms in which the DNA lies free in the cell without being confined in a nucleus. Included are the bacteria and the blue-green algae. Prokaryotes contain no histones and have no nucleosomes.

PROTEIN A substance composed of one or more polypeptides.

RECESSIVE ALLELE One that does not express itself in observable characters unless it occurs in both versions of the genetic information (see Homozygous).

REPETITIVE SEQUENCES See Sequences.

RIBOSOME A tiny structure (organelle) constituting the central element of the protein-synthesizing machinery of the cell. In form the ribosome consists of two unequal hemispheres joined at their flat surfaces. Both hemispheres are composed of RNA and several kinds of protein (see fig. 5).

RNA Ribonucleic acid. A substance essential to life composed of ribonucleotides joined side by side. The sequence of bases in the ribonucleotides of any RNA molecule is determined by the base sequence in some specific portion of the DNA from which that RNA molecule was transcribed. RNA is classified according to its function into several forms. Among them are n-RNA (nuclear RNA), m-RNA (messenger RNA), r-RNA (ribosomal RNA), and t-RNA (transfer RNA). n-RNA is the precursor of m-RNA and contains intervening and flanking sequences as well as translated sequences. Before leaving the nucleus it is processed as a result of which the intervening and flanking sequences are removed. m-RNA brings the information for constructing polypeptides to the protein-synthesizing machinery. r-RNA along with several proteins makes up the ribosome. t-RNA molecules bring the specific amino acids, one at a time, to the point of contact between the m-RNA and the ribosome during the process of protein synthesis (see fig. 5).

SEQUENCES In eukaryotes some sequences in the DNA exist only once in a haploid complement and are referred to as unique sequences. Others are repeated from tens of thousands to millions of times. These are referred to as repetitive sequences.

The unique sequences are thought to code for polypeptides that result from their translation. They may also be termed translated sequences. The repeated sequences are never translated and may be called untranslated sequences.

The translated sequence that codes for a given polypeptide is not continuous but is interrupted one or more times by untranslated sequences that are the intervening sequences or introns. The separated parts of the translated sequences are the exons. At the ends of a set of exons are other untranslated flanking sequences.

When DNA is transcribed in the nucleus, one long RNA molecule results which includes flanking sequences, exons, and introns. This is nuclear RNA (n-RNA). Before it is transferred to the cytoplasm as m-RNA almost everything but the exons are removed from it by processing.

BIBLIOGRAPHY

BACKGROUND

The following list includes works giving background information in ten of the different fields of study on which the exposition of this book has been based.

1. CYBERNETICS AND INFORMATION THEORY

Elsasser, Walter M. 1958. *The Physical Foundation of Biology*. New York: Pergamon Press

Shannon, Claude E., and Warren Weaver. 1959. *The Mathematical Theory of Communication*. Urbana: University of Illinois Press.

Weiner, Norbert. 1948. *Cybernetics*. New York: John Wiley.

2. DEVELOPMENT

Schneirla, T. C. 1966. Behavioral Development and Comparative Psychology. *Quart. Rev. Biol.* 41:238–302.

Waddington, C. H. 1956. *Principles of Embryology*. New York: Macmillan.

3. GENERAL GENETICS

Levine, Louis. 1980. *Biology of the Gene*. St. Louis: Mosby.

Srb, A. M., R. D. Owen, and R. S. Edgar. 1965. *General Genetics*. San Francisco: Freeman.

4. HUMAN EVOLUTION

Dobzhansky, Th. 1962. *Mankind Evolving*. New Haven: Yale University Press.

Leakey, Richard E., and Roger Lewin. 1978. *People of the Lake*. New York: Doubleday.

Pilbeam, David. 1972. *The Ascent of Man*. New York: Macmillan.

5. HUMAN GENETICS

Vogel, F., and A. G. Motulsky. 1979. *Human Genetics*. Berlin, Heidelberg, New York: Springer-Verlag.

6. HUMAN RACES

Coon, Carleton S. 1965. *The Living Races of Man.* New York: Alfred A. Knopf.

Montagu, Ashley. 1964. *Man's Most Dangerous Myth: The Fallacy of Race.* Cleveland: World Publishing Co.

Montagu, Ashley, ed. 1969. *The Concept of Race.* New York: Collier-Macmillan.

7. INTELLIGENCE

Vernon, Philip E. 1979. *Intelligence: Heredity and Environment.* San Francisco: Freeman.

Willerman, Lee. 1979. *The Psychology of Individual and Group Differences.* San Francisco: Freeman.

8. LANGUAGE

Lenneberg, E. H. 1967. *Biological Foundations of Language.* New York: John Wiley.

9. MOLECULAR BIOLOGY

Stent, Gunther, and R. Calendar. 1978. *Molecular Genetics: An Introductory Narrative.* 2d ed. San Francisco: Freeman.

Suzuki, David T., Anthony J. F. Griffiths, and Richard C. Lewontin. 1981. *An Introduction to Genetic Analysis.* 2d ed. San Francisco: Freeman.

Watson, James D. 1976. *Molecular Biology of the Gene.* 3d ed. Menlo Park, Calif.: Benjamin.

10. SPECIATION

Mayr, Ernst. 1963. *Animal Species and Evolution.* Cambridge, Mass.: Harvard University Press.

REFERENCES

Adams, Charles Francis. 1892. *Three Episodes of Massachusetts History*. 2 vols. Boston: Houghton-Mifflin.

Allison, A. C. 1954. Protection Afforded by Sickle-Cell Trait against Subtertian Malarial Infection. *Brit. Med. J.* 1:290–94.

American Ornithologists Union. 1957. *Check List of North American Birds*. 5th ed. Ithaca, N. Y.: Cornell University Press.

———. 1973. 32d supplement to Check List of 1957. *Auk* 90:411–19.

Aristotle. 1962. *Politics*. Translated by T. A. Sinclair. Baltimore: Penguin Books.

Auel, Jean M. 1980. *The Clan of the Cave Bear*. New York: Crown Publishers.

Baker, John R. 1974. *Race*. New York: Oxford University Press.

Bank, Arthur, J. G. Mears, and F. Ramirez. 1980. Disorders of Human Hemoglobin. *Science* 207:486–93.

Benedict, Ruth. 1934. *Patterns of Culture*. Boston: Houghton-Miffllin.

Birch, H. G. 1956. Sources of Order in Maternal Behavior of Animals. *Am. J. Orthopsychiatry* 26:279–84.

Bernstein, B. B. 1971. *Class, Codes, and Control*. London: Routledge and Kegan Paul.

Blumenbach, Johann F. 1795. De generis humani varietate nativa (On the Natural Varieties of Mankind). In *Anthropoligical Treatises*. Translated by Thomas Bendyshe (1865), pp. 145–276. Published for the Anthropological Society. London: Longman, Green, Green, Longman Roberts and Green.

Bodmer, W. F., and L. L. Cavalli-Sforza. 1976. *Genetics, Evolution, and Man*. San Francisco: Freeman.

Boyd, William. 1950. *Genetics and the Races of Man*. Boston: Little Brown.

Brace, C. Loring. 1969. A Nonracial Approach toward the Understanding of Human Diversity. In *The Concept of Race*. Ashley Montagu, ed., pp. 103–52. London: Collier-Macmillan.

———. 1979. *The Stages of Human Evolution*. Englewood Cliffs, N. J.: Prentice-Hall.

Britten, R. J., and D. E. Kohne. 1968. Repeated Sequences in DNA. *Science* 161:529.

Brown, M. S., and J. L. Goldstein. 1976. Receptor-Mediated Control of Cholesterol Metabolism. *Science* 191:150—54.

Brues, Alice M. 1954. Selection and Polymorphism in the ABO Blood Groups. *Am. J. Phys. Anthropol.* n.s. 12:559—97.

Burt, Sir Cyril. 1955. The Evidence of the Concept of Intelligence. *Brit. J. Educ. Psychol.* 25:158—77.

————. 1958. The Inheritance of Mental Ability. 1957 Bingham Lecture. *Am. Psychol.* 13:1—15.

————. 1966. The Genetic Determination of Differences in Intelligence: A Study of Monozygotic Twins Reared Together and Apart. *Brit. J. Psychol.* 57:137—53.

Coon, Carleton S. 1962. *The Origin of Races.* New York: Alfred A. Knopf.

————. 1965. *The Living Races of Man.* New York: Alfred A. Knopf.

Coon, Carleton S., S. M. Garn, and J. P. Birdsell. 1950. *Races.* Springfield, Ill.: C. C. Thomas.

Counce, S. J. 1956. Studies on Female Sterility Genes in *Drosophila melanogaster. Z. Abstamm. und Vererb.* 87:443—92.

Crick, Francis. 1979. Split Genes and RNA Splicing. *Science* 204:264—71.

Darlington, C. D. 1947. The Genetic Component of Language. *Heredity* 1:269—86.

Darwin, Charles. 1859. *On the Origin of Species.* Facsimile edition 1964; Ernst Mayr, ed. Cambridge, Mass.: Harvard University Press.

————. 1871. *The Descent of Man.* London: Murray.

Davenport, C. B., and Mary T. Scudder. 1919. Naval Officers: Their Heredity and Development. Washington, D. C.: Carnegie Institution, Publication 259.

Deutsch, Martin. 1965. The Role of Social Class in Language Development and Cognition. *Am. J. Orthopsychiatry* 35:78—88.

Diamond, J. M. 1978. The Tasmanians: The Longest Isolation, the Simplest Technology. *Nature* 273:185—86.

Disraeli, Benjamin. 1934. *Sybil, or the Two Nations.* World's Classic Edition. London: Oxford University Press.

Dobzhansky, Th. 1962. *Mankind Evolving.* New Haven: Yale University Press.

Dobzhansky, Th., and O. Pavlovsky. 1957. An Experimental Study of Interaction between Genetic Drift and Natural Selection. *Evolution* 11:311—19.

Dobzhansky, Th., and B. Spassky. 1967. Effects of Selection and Migration on Geotactic and Phototactic Behavior of *Drosophila* I. *Proc. Roy. Soc. London Ser. B* 168:27—47.

Dobzhansky, Th., B. Spassky, and J. Sved. 1969. Effects of Selection and Migration on Geotactic and Phototactic Behavior of *Drosophila* II. *Proc. Roy. Soc. London Ser. B* 173:191—207.

Down, A. Langdon H. 1866. Observations on an Ethnic Classification of Idiots. *Clinical Lectures and Reports: London Hospital* 3:259−62.

Dubos, René. 1980. Nutritional Ambiguities. *Natural History* 89:14−21.

Duberman, Martin. 1964. *In White America*. New York: New American Library.

Eichenwald, H. F., and P. C. Fry. 1969. Nutrition and Learning. *Science* 163:644−48.

Eliot, George. 1910. *Romola*. Boston: Dana Estes.

Erlenmeyer-Kimling, L., and L. F. Jarvik. 1963. Genetics and Intelligence: A Review. *Science* 142:1477−78.

Feldman, M. W., and R. C. Lewontin. 1975. The Heritability Hang-up. *Science* 190:1163−68.

Felsenfeld, G. 1978. Chromatin. *Nature* 271:115−22.

Fischer, E. 1913. *Die Rehobother Bastards und das Bastardierungsproblem beim Menschen*. Jena: Gustav Fischer Verlag.

Fisher, Vardis. 1944. *The Golden Rooms*. New York: Vanguard Press.

Fox, Robin. 1971. The Cultural Animal. In *Man and Beast: Comparative Social Behavior*. Eisenberg, J. F., and W. Dillon, eds., pp. 273−96. Washington, D.C.: Smithsonian Institution Press.

Fredrickson, D. S., J. L. Goldstein, and M. S. Brown. 1978. The Familial Hyperlipoproteinemias. In *The Metabolic Basis of Inherited Disease*. 4th ed., Stanbury, J. B., J. B. Wyngaarden, and D. S. Fredrickson, eds., chap. 30, pp. 604−655. New York: McGraw-Hill.

Friedman, Milton J., and William Trager. 1981. The Biochemistry of Resistance to Malaria. *Scientific American* 244 (Feb.): 154−164.

Galsworthy, John. 1921. *To Let*. New York: Charles Scribner's Sons.

Galton, Francis. 1962. *Hereditary Genius*. Cleveland: World Publishing Co.

Geist, Valerius. 1981. Neanderthal the Hunter. *Natural History* 90 (Jan.): 26−36.

Gillie, O. 1976. Crucial Data was Faked by Eminent Psychologist. *Sunday Times* (London) Oct. 24, 1976.

Gladwin, Thomas. 1970. *East is a Big Bird*. Cambridge, Mass.: Harvard University Press.

Glass, Bentley. 1956. Genetic Drift in Human Populations. *Am. J. Phys. Anthropol.* n.s. 14:541−55.

de Gobineau, J. A. 1853, 1855. *Essai sur l'inégalité des races humaines*. Paris: Librairie de Firmin Didot Frères.

─────. 1856. *The Moral and Intellectual Diversity of Races*. Philadelphia: J. B. Lippincott.

Golding, William. 1955. *The Inheritors*. New York: Harcourt Brace World.

Griffin, James E. 1979. Testicular Feminization Associated with a Thermolabile Androgen Receptor in Cultured Human Fibroblasts. *J. Clin. Inv.* 64:1624–31

Grosvenor, Edwin A. 1918. The Races of Europe. *National Geographic Magazine* 34:441–532.

Guilford, J. P. 1967. *The Nature of Human Intelligence.* New York: McGraw-Hill.

Hamilton, W. D. 1964. The Genetical Theory of Social Behavior I, II. *J. of Theoretical Biol.* 7:1–52.

Hadorn, Ernst. 1961. *Developmental Genetics and Lethal Factors.* New York: John Wiley.

Hearnshaw, L. S. 1979. *Cyril Burt, Psychologist.* Ithaca, N. Y.: Cornell University Press.

Holt, Sarah B. 1961. Quantitative Genetics of Fingerprint Patterns. *British Medical Bulletin* 17:247–50.

Hooton, Earnest A., and C. W. Dupertuis. 1955. The Physical Anthropology of Ireland. Papers of the Peabody Museum of Archeology and Ethnology, vol. 30, nos. 1 & 2. Cambridge, Mass.: Harvard University, published by the Museum.

Howell, F. Clark, and the Editors of Time-Life Books. 1968. *Early Man.* New York: Time-Life Books.

Howells, W. 1954. *Back of History.* Garden City, N. Y.: Doubleday.

Hunt, J. M. 1961. *Intelligence and Experience.* New York: The Ronald Press.

———. 1969. Has Compensatory Education Failed? Has it been Attempted? *Harvard Educ. Rev.* 39:130–52.

Huxley, Thomas H. 1910. *Lectures and Lay Sermons.* New York: E. P. Dutton.

Jencks, Christopher. 1972. *Inequality: A Reassessment of the Effect of Family and Schooling in America.* New York: Basic Books.

Jensen, Arthur R. 1969*a*. How Much Can We Boost IQ and Scholastic Achievement? *Harvard Educ. Rev.* 39:1–123.

———. 1969*b*. Reducing the Heredity-Environment Uncertainty. *Harvard Educ. Rev.* 39:209–43.

———. 1970. IQs of Identical Twins Reared Apart. *Behavior Genetics* 1:133–48.

———. 1972. *Genetics and Education.* New York: Harper and Row.

———. 1974. Kinship Correlations Reported by Sir Cyril Burt. *Behavior Genetics* 4:1–28.

———. 1980. *Bias in Mental Testing.* New York: The Free Press.

Jinks, L. J., and D. W. Fulker. 1970. Comparison of the Biometrical, Genetical, MAVA, and Classical Approaches to the Analysis of Human Behavior. *Psychol. Bull.* 73:311–49.

Johnston, R. F., and R. K. Selander. 1964. House Sparrows: Rapid Evolution of Races in North America. *Science* 144:548–50.

Kagan, Jerome. 1973. Cross-cultural Perspectives on Early Development. *Am. Psychol.* 28:947−61.

Kamin, L. J. 1974. *The Science and Politics of I. Q.* Potomac, Md.: Lawrence Erlbaum Associates.

————. 1977. Burt's IQ Data (Letter to *Science*). *Science* 195:246−48.

Keenan, B. S., W. J. Meyer III, A. J. Hadjian, H. W. Jones, and C. J. Midgeon. 1974. Syndrome of Androgen Sensitivity in Man: Absence of 5-Alpha-Dihydrotestosterone Binding Protein in Skin Fibroblasts. *J. Clin. Endocrinol. Metab.* 38:1143−46.

Kimara, Kunihiko. 1967. A Consideration of the Secular Trend in Japanese Height and Weight by a Graphic Method. *Am J. Phys. Anthropol.* 29:89−94.

King, James C. 1961. Inbreeding, Heterosis, and Information Theory. *American Naturalist* 95:345−64.

King, James C., and Lauritz Sømme. 1958. Chromosomal Analyses of the Genetic Factors for Resistance to DDT in Two Resistant Lines of *Drosophila melanogaster. Genetics* 43:577−93.

King, Mary Claire, and A. C. Wilson. 1975. Evolution at Two Levels in Humans and Chimpanzees. *Science* 188:107−16.

Kurtén, Björn. 1980. *Dance of the Tiger.* New York: Pantheon Books.

Latter, B. D. H. 1980. Genetic Differences within and between Populations of the Major Human Subgroups. *American Naturalist* 116:220−37.

Layzer, David. 1974. Heritability Analyses of I.Q. Scores: Science or Numerology? *Science* 183:1259−66.

Leakey, Richard E., and Roger Lewin. 1978. *The People of the Lake.* New York: Doubleday.

Leakey, Richard E., and Alan C. Walker. 1976. *Australopithecus, Homo erectus,* and the Single Species Hypothesis. *Nature* 261:572−74.

Leder, Philip. 1978. Discontinuous Genes. *New Eng. J. Med.* 298:1079−81.

Lee, Richard B. 1979. *The !Kung San: Men, Women, and Work in a Foraging Society.* New York: Cambridge University Press.

Lejeune, J., M. Gautier, and R. Turpin. 1959. Etude des chromosomes somatiques de neuf enfants mongoliens. *Compt. Rend. Acad. Sci.* (Paris) 248:1721−22.

Lerner, I. M. 1954. *Genetic Homeostasis.* New York: John Wiley.

————. 1972. Polygenic Inheritance and Human Intelligence. *Evolutionary Biol.* 6:399−414.

Lerner, I. M., and W. J. Libby. 1976. *Heredity, Evolution, and Society.* 2d ed. San Francisco: Freeman.

Lewontin, R. C. 1972. The Apportionment of Human Diversity. *Evolutionary Biol.* 6:381−98.

Livingstone, Frank B. 1967. *Abnormal Hemoglobins in Human Populations.* Chicago: Aldine.

Lorenz, Konrad. 1965. *Evolution and the Modification of Behavior.* Chicago: University of Chicago Press.

————. 1966. *On Aggression.* Translated by Marjorie Kerr Wilson. New York: Harcourt Brace.

Maynard Smith, John. 1975. Molecular Evolution and the Age of Man. *Nature* 253:497−98.

Mayr, Ernst. 1954. Changes of Genetic Environment and Evolution. In *Evolution as a Process*, J. Huxley, A. C. Hardy, and E. B. Ford, eds., pp. 157−80. London: Allen and Unwin.

————. 1955. The Species as a Systematic and as a Biological Problem. *Biological Systematics: Proceedings of the Sixteenth Annual Biological Colloquium.* Corvallis, Oreg.: Oregon State College.

————. 1963. *Animal Species and Evolution.* Cambridge, Mass.: Harvard University Press.

Meyer, W. J. III, B. R. Midgeon, and C. J. Midgeon. 1975. Locus on Human X-Chromosome for Dihydrotestosterone Receptor and Androgen Insensitivity. *Proc. Nat. Acad. Sci. U. S.* 72:1469−72.

Miller, A. H. 1941. Speciation in the Avian Genus *Junco. Univ. Calif. Publ. Zool.* 44:173−434.

Mitford, Jessica. 1960. *Daughters and Rebels.* Boston: Houghton-Mifflin.

Montagu, Ashley. 1964. *Man's Most Dangerous Myth: The Fallacy of Race.* Cleveland: World Publishing Co.

————. ed. 1969. *The Concept of Race.* London: Collier-Macmillan.

————. ed. 1980. *Sociobiology Examined.* New York: Oxford University Press.

Morton, N. E. 1962. Genetics of Interracial Crosses in Hawaii. *Eugenics Quarterly* 9:23−24.

Muir, D. W., and D. E. Mitchell. 1973. Visual Resolution and Experience: Acuity Deficits in Cats following Early Selective Deprivation. *Science* 180:420−22.

Myrianthopoulos, N. C., and S. M. Aronson. 1966. Population Dynamics of Tay-Sachs Disease I. Reproductive Fitness and Selection. *Am. J. Human Genet.* 18:313−27.

Nei, M., and A. K. Roychoudhury. 1974. Genetic Variation within and between the Three Major Races of Man, Caucasoids, Negroids, and Mongoloids. *Am. J. Human Genet* 26:421−43.

Nicolson, Sir Harold George. 1934. *Curzon: The Last Phase.* New York: Harcourt Brace.

O'Malley, Bert W., and W. T. Schrader. 1976. The Receptors of Steroid Hormones. *Scientific American* 234 (Feb.):32−43.

Orgel, L. E., and F. H. C. Crick. 1980. Selfish DNA: The Ultimate Parasite. *Nature* 284:604−607.

Pilbeam, David. 1972. *The Ascent of Man*. New York: Macmillan.

Plato. 1955. *The Republic*. Translated by H. D. P. Lee. Baltimore: Penguin Books.

Proudfoot, Nicholas J., Monica H. M. Shander, Jim L. Manley, Malcolm L. Gefter, and Tom Maniatis. 1980. Structure and in vitro Transcription of Human Globin Genes. *Science* 209:1329—36.

Reed, Sheldon. 1963. *Counseling in Medical Genetics*. Philadelphia: Saunders.

Reed, T. E. 1969. Caucasian Genes in American Negroes. *Science* 165:762—68.

Ripley, W. Z. 1899. *The Races of Europe*. New York: D. Appleton.

Sahlins, Marshall. 1977. *The Use and Abuse of Biology*. Ann Arbor: University of Michigan Press.

Savage-Rumbaugh, E. Sue, Duane M. Rumbaugh, and Sarah Boysen. 1980. Do Apes Use Language? *American Scientist* 68:49—61.

Schrire, Carmel. 1980. Hunter-Gatherers in Africa. *Science* 210:890—91.

Schmid, Carl W., and P. L. Deininger. 1975. Sequence Organization of the Human Genome. *Cell* 6:345—58.

Schneirla, T. C. 1966. Behavioral Development and Comparative Psychology. *Quart. Rev. Biol.* 41:283—302.

Shapiro, H. L. 1929. *Descendants of the Mutineers of the Bounty*. Honolulu: Mem. Bishop Museum, 2:1—106.

Shaw, R. F., and A. P. Smith. 1969. Is Tay-Sachs Disease Increasing? *Nature* 224:1214—15.

Stein, G. S., J. C. Spelsberg, and L. J. Kleinsmith. 1974. Nonhistone Chromosomal Proteins and Gene Regulation. *Science* 183:817—24.

Stern, C., and E. W. Schaeffer. 1943. On Wild-Type Isoalleles in *Drosophila melanogaster*, *Proc. Nat. Acad. Sci. U. S.* 29:361—67.

Terrace, H. S., L. A. Pettito, R. J. Sanders, and T. G. Bever. 1979. Can an Ape Create a Sentence? *Science* 206:891—902.

Trinkaus, Erik, and William W. Howells. 1979. The Neanderthals. *Scientific American* 241 (Dec) 118—133.

Trivers, R. L. 1971. The Evolution of Reciprocal Altruism. *Quart. Rev. Biol.* 46:35—57.

———. 1972. Parental Investment and Sexual Selection. In *Sexual selection and the descent of man, 1871—1971*. B. G. Campbell, ed., pp. 136—179. Chicago: Aldine Press.

———. 1974. Parent-Offspring Conflict. *American Zoologist* 14:249—64.

Tuan, Dorothy, M. J. Murnane, J. K. deRiel, and B. G. Forget. 1980. Heterogeneity in the Molecular Basis of Hereditary Persistence of Fetal Hemoglobin. *Nature* 285:335—37.

Ullrich, Axel, T. J. Dull, A. Gray, J. Brosius, and Irmi Sures. 1980. Genetic Variation in the Human Insulin Gene. *Science* 209:612—15.

United States Public Health Service. 1965. Publication 1000, no. 8. Vital and Health Statistics. Washington, D. C.: Government Printing Office.

Van der Kloot, W. G., and C. M. Williams. 1953. Cocoon Construction by the Cecropia Silkworm. I. The Role of External Environment. *Behavior* 5:141−56.

Van der Ploeg, A. K., M. Oort, D. Roos, L. Bernini, and R. A. Flavell. 1980. Gamma-Beta Thalassemia Studies Showing that Deletion of the Gamma- and Delta-Genes Influences Beta-Globin Gene Expression in Man. *Nature* 283:637−42.

Vernon, P. E. 1979. *Intelligence: Heredity and Environment*. San Francisco: Freeman.

Waddington, C. H. 1953. Genetic Assimilation of an Acquired Character. *Evolution* 7:118−26.

Wade, N. 1976. Sociobiology: Troubled Birth for a New Discipline. *Science* 191:1151−55.

Walker, Alan, and R. E. F. Leakey. 1978. The Hominids of East Turkana. *Scientific American* 239 (Aug.):54−66.

Watson, J. D. 1976. *Molecular Biology of the Gene*. 3d ed. Menlo Park, Calif.: Benjamin.

Wechsler, David. 1955. *Manual for the Wechsler Adult Intelligence Scale*. New York: The Psychological Corporation.

Willerman, Lee. 1979. *The Psychology of Individual and Group Differences*. San Francisco: Freeman.

Wilson, E. O. 1975. *Sociobiology: The New Synthesis*. Cambridge, Mass.: Belknap Press of Harvard University Press.

———. 1978. *On Human Nature*. Cambridge, Mass.: Harvard University Press.

Wilson, J. D., and P. C. MacDonald. 1978. Male Pseudohermaphroditism due to Androgen Resistance: Testicular Feminization and Related Syndromes. In *The Metabolic Basis of Inherited Disease*. Stanbury, J. B., J. B. Wyngaarden, and D. S. Fredrickson, eds., chap. 42, pp. 894−913. New York: McGraw-Hill.

Winick, Myron, K. K. Meyer, and R. C. Harris. 1975. Malnutrition and Environmental Enrichment by Early Adoption. *Science* 190:1173−75.

Yanofsky, C., V. Horn, and D. Thorp. 1964. Protein Structure Relationships Revealed by Mutational Analysis. *Science* 146:1593−94.

Yunis, J. J., J. R. Sawyer, and K. Dunham. 1980. The Striking Resemblance of High-Resolution G-Banded Chromosomes of Man and Chimpanzee. *Science* 208:1145−48.

Zajonc, R. B. 1976. Family Configuration and Intelligence. *Science* 192:227−36.

INDEX

Adams, Charles Francis, Jr., 119
Adaptiveness: of face form, 142; of geographic variation, 139—146; of human phenotypic variation, 138—146; of nose form, 141—142; of pigmentation, 140—141
Albino, human, 21
Alleles, defective, 97, 101; interaction between, 92—95; spectrum of multiple, 51—59. *See also* Isoalleles
Allison, A. C., 145
Allopatric species, 18
Alpine race, 111
American (Indian) race, 10, 111
Amino acid, 35—36
Apes, relation to *Homo sapiens*, 126—130
Aristotle, 67, 117
Aronson, S. M., 144
Aryan, language and race, 113—114, 118
Ashkenazim, 143—144
Auel, Jean M., 132
Australopithecus africanus, A. robustus, 131

Bacteria, function by means of interlocking feedbacks, 22, 39—41
Baker, John R., 133
Baltic race, 111
Bank, Arthur, 56, 57, 59
Basques, 114—115
Behavior: as phenotype, genetic influences on, 61—67; innate, 64; species-specific, 64, 66; species-typical, 66
Behavioral characters, selection for in *Drosophila*, 121—122
Benedict, Ruth, 81
Bernini, L., 56, 57
Bernstein, B. B., 81
Bever, T. G., 68
Biological species, 4—5
Biotic program, 14
Birch, H. G., 62
Birdsell, J. P., 141, 157
Bits: as units of information, 22; number of in genetic complements of *E. coli* and *H.*

sapiens, 23
Blacks: measure of white alleles in American, 146—147; status in United States and Spanish Caribbean, 155; status in Britain, 156
Blood groups: ABO, 102—104; and racial history, 149—150; MN, 106; Rh, 114—115; as evidence of migration, 149—150
Blumenbach, Johann F., 10, 111, 123, 125—126, 127, 157
Bodmer, W. F., 144
Boyd, William, 149
Boysen, Sarah, 68
Brace, C. Loring, 138, 141
Brachyrury, T, as mutant in mice, 49
Breeds of domestic animals, 6—7
Britten, R. J., 43
Brosius, J., 57
Brown, M. S., 45, 46
Burt, Sir Cyril, 73; controversy over published work, 83—85
Bushmen, 69, 85, 140

Caste system, India, 135
Cat, domestic: behavior of kittens and mother, 66; effect of light stimuli on developing vision, 76
Cattell, J. M., 83
Cattell, R. B., 83
Caucasian race, 10, 111
Cavalli-Sforza, L. L., 144
Chicken combs, 92—94
Chimpanzee: chromosomes compared with *H. sapiens*, 128; light deprivation, influence on vision, 76; linguistic competence, 68; phenotype, relation to human races, 129; proteins compared with *H. sapiens*, 129
Cholesterol: regulation of synthesis in humans, 45—46
Chromosomes: 42, 92; aberrations, 97—98; comparison of human and chimpanzee, 128
Circular overlap, 18
Classes, human socioeconomic: diagnostic characteristics, 119—120; differences in I.Q.,

175

Designer: Graphics Two
Compositor: Trend Western
Printer: Vail-Ballou
Binder: Vail-Ballou
Text: 10/12 Garamond set on Linoterm
Display: Garamond